THE CAMBRIDGE SERIES OF PHYSICAL CHEMISTRY

GENERAL EDITOR

E. K. RIDEAL

Professor of Colloid Science in the
University of Cambridge

MOLECULAR RAYS

MOLECULAR RAYS

By

RONALD G. J. FRASER

PH.D. (Aberd.)

CAMBRIDGE

AT THE UNIVERSITY PRESS

1931

CAMBRIDGE
UNIVERSITY PRESS

University Printing House, Cambridge CB2 8BS, United Kingdom

Cambridge University Press is part of the University of Cambridge.

It furthers the University's mission by disseminating knowledge in the pursuit of education, learning and research at the highest international levels of excellence.

www.cambridge.org
Information on this title: www.cambridge.org/9781107593411

© Cambridge University Press 1931

First published 1931
First paperback edition 2015

A catalogue record for this publication is available from the British Library

ISBN 978-1-107-59341-1 Paperback

PREFACE

I have agreed very gladly to Professor Rideal's request to write a monograph on Molecular Rays for the Cambridge Series of Physical Chemistry, because I believe that a connected account of this fascinating study will serve at once to interest an increasingly large company in the experimental results which have been obtained with the method, and to serve as a general guide to the technique to those who themselves desire to use it as a tool in research.

The book has been written strictly from the experimental standpoint, and is intended primarily to give a balanced survey of the whole field rather than a minute examination of its separate parts. Chapter I, which deals with the technique of the production and measurement of the rays has however been given rather more detail than the rest of the book, and contains a considerable body of laboratory experience which was inevitably omitted from the original papers; and it is hoped that this part of the book may consequently be useful not only as a general account of the technical side of the subject but also in a measure as a laboratory manual for those who, without previous experience, are entering on experimental work in Molecular Rays.

The literature up to the end of 1930, with the exception of a few deliberate omissions, has been as far as possible completely covered.

I have very many acknowledgments to render to those who have helped me to make this book. Above all, I have to thank Professor Stern and all those friends, both German and American, with whom I worked in the Hamburg Institute for Physical Chemistry, for help given in the form of unpublished data, original photographs for reproduction, line drawings, and some outspoken criticism. This book contains much of real value, however, which is due to no single individual, but which has been taken from the common stock of opinion and experience current in Stern's laboratory at Hamburg. To that more intangible source also I would acknowledge my debt.

I have to thank Dr P. Clausing of Eindhoven and Dr T. H. Johnson of the Bartol Research Foundation, in each case for a valuable correspondence: and the latter particularly for the very beautiful pictures reproduced in Plate III; also Dr L. C. Jackson for permission to include an account of preliminary results on active nitrogen, Dr J. D. Cockcroft for the original of Fig. 29, Dr O. E. Kurt for the original of Fig. 51, and the Royal Commissioners for the Exhibition of 1851 for the loan of the originals from which the enlargements of Plate VII were made.

My particular thanks are due to Dr A. Klinkenberg and Dr L. F. Broadway for carrying out a number of laborious calculations at my request, and to Dr Broadway for checking all numerical calculations. Mr S. A. McKay has given me great assistance in the preparation of line drawings and graphs. Dr F. P. Bowden and Dr C. P. Snow have read parts of the manuscript, and have made many valuable suggestions, not always adopted, towards its betterment. Dr W. H. Watson has gallantly read the entire book in proof, and his merciless exposure of the traditional "obscurities in the text" has left me deeply in his debt.

Ewald's *Kristalle und Röntgenstrahlen*, Berlin, 1923 has been of great assistance to me pedagogically in writing Chapter IV; those who have read Ewald's book will realise how much the presentation of parts of that chapter owes to it.

I have to thank the Council of the Royal Society, Messrs Julius Springer, Messrs the Akademische Verlagsgesellschaft, *The Physical Review*, *Physica* for the use of line and half-tone blocks; and the McGraw Hill Book Company for permission to reproduce Table VI, p. 123, from Pauling and Goudsmit's *Structure of Line Spectra.*

It is with particular pleasure that I record my indebtedness to Imperial Chemical Industries Limited, who have made it possible for me financially to pursue my experimental work and at the same time have the leisure to indulge in the writing of books.

RONALD FRASER

CAMBRIDGE
August 1931

FOREWORD

The investigations which are pursued in the Hamburg Institute for Physical Chemistry, and in which Dr Fraser took part for three years, are concerned with the task of furthering the development of the physical method of Molecular Rays. This task embraces not only the evolution of the experimental technique, but also the exploitation of the characteristic features of the new method, the selection of problems which lend themselves to its attack.

Since Dr Fraser discusses in his book chiefly the experimental aspect of the subject, I would add here some general remarks concerning the second point above.

What are the characteristic features of the molecular ray method? Its directness and (in principle at least) its primitiveness. These characteristics at the same time define the range of the method. That is clearly shown in its historical development.

The method was used initially to verify the fundamental postulates of the gas kinetic theory. Naturally there was never any doubt about their validity; but it was none the less satisfactory that one was able to demonstrate so absolutely directly the linear motion of the molecules, to measure their velocity, and so forth.

The characteristics of the method are still more clearly shown in the next problem attacked, that of so-called space quantisation. The quantum theory required that magnetic atoms should take up only certain discrete positions in a magnetic field. Consequently one was at that time (1921) forced to conclude that a gas composed of such atoms should show marked double refraction in a magnetic field. Experiment demonstrated not the slightest trace of this double refraction, a contradiction which in the then stage of the theory was incomprehensible. The molecular ray method made possible an experimentia crucis. It gave absolutely direct evidence of the discrete positions (or rather components of the magnetic moment) demanded by the quantum theory. The fact that the method yields directly the terms (energy values in the field) and not term differences as in the optical method, was in this instance particularly important. It is clear that it is precisely the directness and (essential) primitiveness of the molecular ray method which enable it to attack problems of a fundamental character. The latest developments show the same trend. It has been found possible to diffract beams of atoms and molecules at the cross grating of a crystal cleavage plane, and thus to establish directly the wave properties demanded by the new mechanics.

Finally, the characteristics already mentioned are evidenced in a further wide field of enquiry open to the molecular ray method, namely the investigation of molecular properties. If one wishes to investigate the magnetic or electric properties of molecules, then the most obvious and natural procedure is surely this: One takes a beam of molecules, sends it past the pole of a magnet or an electrically charged body, and watches how the molecules are deflected. In the investigation of molecular properties appear as further characteristics of the method its high sensitivity (nuclear moments), and its suitability for the study of the potential fields around molecules and at surfaces, whereby the wave mechanics, with its connection between potential and refractive index of the matter waves, plays a special part.

I hope that I have been successful, as a result of this description, in very rough outline, of the characteristics of the molecular ray method and of its most important problems, in indicating also something of that beauty and peculiar charm which so firmly captivate physicists working in this field. I believe that the new method will in time find far wider application than heretofore, and I hope that the present book will help to further this development.

O. STERN

HAMBURG
July 1931

CONTENTS

LIST OF PLATES

INTRODUCTION

The fundamental postulate of the kinetic theory of gases is that of molecular chaos. According to the theory, a quantity of gas consists of a very large number of discrete particles, or molecules, the motion of which is incessant and entirely uncoordinated. In other words, the molecules of a gas do not describe regular orbits, but follow completely random zig-zag paths, each section of the zig-zag being terminated by a collision, either with another molecule or with a wall of the containing vessel. Between collisions, the molecules describe rectilinear paths with uniform velocity; that is, every section of a zig-zag is strictly a straight line.

The velocity of a specific molecule at any instant is unpredictable; but the velocities of the individual molecules are distributed according to statistical laws about a mean value, which is defined precisely by the external conditions. Corresponding to the mean velocity, there exists for a given molecular concentration a perfectly definite mean path between collisions, or mean free path. Thus the existence of the quantities mean velocity, mean free path, and so forth, which serve to characterise a gas under specified conditions, is dependent on the fact that we are dealing with enormous numbers of molecular individuals, to which large numbers alone statistical laws properly apply. It follows, conversely, that the deduction from the behaviour of a gas of the characteristics of the constituent molecules is only possible with the aid of often complicated statistical considerations.

Now suppose that we pierce the wall of a vessel containing a gas, and allow the molecules to effuse through the aperture so formed into an evacuated space (Fig. 1 a). Then, provided that the pressure in the containing vessel A is not too high, those molecules which happen to approach the dividing wall

between A and the evacuated space B at the position of the
aperture will fly into the space B, and will traverse rectilinear
paths radiating out from the aperture until they hit the walls.
The motion of the molecules in B is already no longer chaotic.

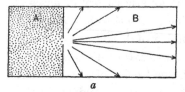
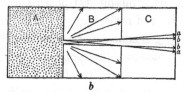

a   b

Fig. 1

If now B is divided into two compartments, B and C, by a
wall pierced by an aperture (Fig. 1 b), only those molecules
will enter C whose directions of motion lie within the boundaries of the cones defined by the apertures in the walls of
A and B. Moreover, their paths will not intersect, if the
aperture in the wall of A is sufficiently small to be regarded
as a point source: thus the only collisions to which the molecules in C will be subject are those arising from the faster
molecules overtaking the slower ones which are traversing
the same path. But this effect is small, and the molecules in
C will be for all practical purposes collision free. The selective
action of the two diaphragms will, in fact, replace the chaotic
motion of the molecules in A by a coordinated motion in C,
where the molecules move in a geometrically defined beam,
or ray. It should therefore be possible with the help of such
a *molecular ray* to approach the study of molecular properties directly, instead of through the channels of statistical
reasoning.

These deductions from gas kinetic theory were brilliantly
justified by the pioneer work of Dunoyer.[a] A cylindrical glass
tube some 20 cm. in length was divided into three compartments by two circular diaphragms of glass. After evacuation,
a little redistilled sodium was introduced into the lowest com-

[a] Dunoyer, *Compt. rend.* 152, 594, 1911; *Le Radium*, 8, 142, 1911.

partment. The sodium was heated to a temperature sufficiently high to vaporise it, when in time there appeared a sharply defined metallic deposit on the upper closed end of the third compartment. The deposit had precisely the size to be expected from the geometry of the apparatus, on the assumption that the sodium atoms describe rectilinear paths between source and deposit. An "umbra" and "penumbra" of the dimensions to be expected if they are defined by the cones bb and aa (see Fig. 1 b) were clearly observable. Objects placed in the path of the beam cast clearly defined "shadows." With an indifferent vacuum in the tube, the deposit was no longer sharp, but diffuse and ill-defined in outline; which is what one must expect as soon as the mean free path of the sodium atoms is shorter than the distance between source and deposit.

From these beginnings in 1911, an elaborate and increasingly important technique has been developed, chiefly at the hands of Stern and his collaborators.[a] The orientation of the work has changed greatly since the time when the experiments were regarded mainly as striking confirmation of the essential accuracy of the conceptions of the dynamical theory of gases. Increasingly wide use has been made of the coordinated character of the motion of the molecules in a beam to investigate directly the properties of the individual molecules; and the molecular ray technique has become a powerful tool in the study of the structure of matter. It is this tool, and some of its applications, which we aim to describe in the following pages.

It should already be sufficiently clear that a molecular ray is a beam of *neutral* molecules, moving *in vacuo* with thermal velocities corresponding to the temperature of the source. The

[a] The main body of this work is contained in a series of papers in the *Zeitschrift für Physik*, from 1926 onwards, under the general title "Untersuchungen zur Molekularstrahlmethode aus dem Institut für physikalische Chemie der Hamburgischen Universität" (cited as U. z. M. 1, etc.).

average velocity of the molecules in a molecular ray lies therefore between some 10^4 cm./sec. and 10^5 cm./sec., depending on the molecular species and on the temperature. A molecular ray can in principle be produced from any stable molecular (including atomic) species whatsoever; the range of the method in this direction is limited solely by the technical possibilities. The method is, however, in general ill-adapted to the study of *excited* molecules; this is at once clear when it is recalled that the time of excitation is of the order 10^{-7} to 10^{-8} sec.; the length of path traversed by a molecule, once excited, before it returns to the normal state, is thus only some 10^{-4} to 10^{-2} cm. This was demonstrated purely experimentally by Dunoyer.[a] He showed that a sodium beam could be rendered visible by exciting it to emit the resonance radiation; when only a central band was irradiated by the D-lines, the boundaries of the luminous portion were seen to be extremely sharp, coinciding as far as could be observed exactly with those of the exciting beam. Although the method is thus not suited to the general study of excited states, the investigation of *metastable* states is within its scope; here the difficulties are the purely technical ones of producing a sufficient concentration of metastable molecules in the source.

A distinction must be drawn between a molecular ray and a *jet* of gas (Dampfstrahl). A molecular ray is practically collision free; within a jet there is turbulent motion. It is therefore impossible for a jet to maintain its form for more than a few millimetres of path *in vacuo*. There is no sharp limit separating the conditions for the production of a molecular ray on the one hand, and a gas jet on the other: the one passes continuously over into the other as the pressure in the source is raised.

The technique is concerned with the production of a molecular ray of the maximum possible intensity. It may appear at first sight, when one considers how few are the essential

---

[a] Dunoyer, *Compt. rend.* **157**, 1068, 1913; *Le Radium*, **10**, 400, 1913.

arrangements necessary to the formation of a molecular ray, that the technique should be one of comparative simplicity. It must be remembered, however, that for many purposes beams of very narrow cross section (ca. 0·01mm.) must be used, and it becomes then a matter of some difficulty to obtain a sufficiently intense beam. Nevertheless, the main details of the production of the rays are now well understood, thanks almost entirely to the work of the Hamburg school.[a]

There remains the problem of detection. The most familiar form of detector is of course the primitive cooled target, on which the beam marks in time the imprint of its cross section; the target detector is, however, only semi-quantitative, and by no means of universal application. There are indeed no perfectly general methods of detection of the molecular rays, such as exist for the detection of charged particles of high energy; each molecular species has to be considered on its merits and a suitable detector devised for use with it. Several satisfactory types of quantitative detectors, of more or less limited range of application, have been developed; but the great majority of substances await the advent of suitable metrical devices for their completely successful study. The central problem on the technical side is the problem of detection. (Chapter 1.)

The most immediate use of a perfected technique of molecular rays lies in the direct experimental study of gases. Here much has already been accomplished. The central law of the kinetic theory of gases, the Maxwell Law of Distribution of Velocities, has been directly confirmed. An elegant method of measuring the mean free path has been developed. A wide field of study, that of intermolecular scattering, has been opened up, and important information about the nature of

---

[a] The fundamental principles of the production of intense molecular beams of narrow cross section were laid down in 1926 by Stern (*Z. Physik*, **39**, 751, 1926. U. z. M. 1).

molecular fields may be expected to accrue from future investigations in this direction. On the technical side, it has been found possible to produce molecular beams of sensibly uniform velocity, which should prove extremely valuable for many applications of the molecular ray method (Chapter 2). It should be possible to observe directly the dissociation of diatomic molecules by means of the velocity selectors which are used to produce beams of uniform velocity (Chapter 7).

Intensive study of the mechanism of collision and condensation of molecules at solid surfaces has been made possible by the advances in vacuum technique which have been made during the last twenty years. The ability to confine the impinging molecules within the definite limits of a directed beam is clearly often of great value, and the molecular ray method has already yielded results of fundamental importance in the investigation of the scattering and reflection of molecules at solid surfaces.

The most recent work in this field has been largely guided, and considerably clarified, by the recognition of the wave nature of matter. It will be recalled that de Broglie[a] suggested that just as light displayed a dual character, undulatory and corpuscular, so matter should likewise be regarded as possessing a dual character, corpuscular and undulatory. It was known that the momentum of a light quant, frequency ν, wavelength λ, could be expressed as

$$\frac{h\nu}{c} = \frac{h}{\lambda},$$

where c is the velocity of light, and h is Planck's constant. de Broglie put forward the hypothesis that the momentum mv of a moving particle could be written

$$mv = \frac{h}{\lambda},$$

where λ is the wavelength of the plane wave which represents

[a] de Broglie, *Phil. Mag.* 47, 446, 1924.

the rectilinear motion of the particle. He supported the hypothesis by a fine mathematical analysis.

The hypothesis was brilliantly confirmed, for the electron, by Davisson and Germer in America, and by G. P. Thomson in this country; and electron diffraction has become an intensively cultivated field of research.[a] It is in some ways surprising that the effect was not discovered earlier, purely experimentally; but when it is remembered that the de Broglie wavelength for electrons of say 25,000 volts energy is only 0.75×10^{-9} cm., it is clear that only a lucky chance could have disclosed the existence of such short waves without the guidance of theory.

Now it is an essential feature of de Broglie's theory that the rectilinear motion of *every* moving mass can be represented by a plane wave. Thus the de Broglie wavelength of for example hydrogen at $0°$ C. is of the order

$$\lambda = \frac{h}{mv} \sim \frac{6.54 \times 10^{-27}}{3.3 \times 10^{-24}.1.7 \times 10^5} \sim 1.2 \times 10^{-8} \text{ cm.},$$

that is, of the same order as that of X rays. For heavier elements and higher temperatures the de Broglie wavelength of the molecules becomes extremely short, and difficult of observation. Attention has therefore been confined hitherto chiefly to the lightest gases.

It has been found that certain of the conditions for the specular reflection of a molecular ray at a solid surface appear to be just those for the reflection of light from a matt surface (Chapter 3); and further, that beams of hydrogen, helium, and atomic hydrogen reflected from the cleavage surface of a crystal are split up into diffraction patterns, which have precisely the characteristics demanded of the reflection of plane waves of the predicted de Broglie wavelength from the lattice formed by the regularly spaced ions in the crystal face (Chapter 4).

[a] For references to the literature, see G. P. Thomson, *Wave Mechanics of Free Electrons*, New York, 1930.

On a more prosaic level, the study of metal films deposited on solid surfaces from molecular beams has given a better insight into the process of the formation and growth of condensates, and an indication of the conditions under which they may adequately be observed (Chapter 3).

The most intensive application of the molecular ray technique to a particular problem has lain until recently in the investigation of the magnetic and electric properties of molecules. The method, which was initiated by the classic experiments of Stern and Gerlach in 1922, depends on the fact that a beam of atoms or molecules which possess a magnetic or electric moment suffers a deviation when shot through an inhomogeneous magnetic or electric field. The extent and character of the deviation can be used to evaluate the magnitude of the moment, and to yield valuable information about the energy states of the molecules constituting the beam (Chapters 5 and 6). The method is of peculiar value in that it serves to measure directly the effect of the field on isolated molecules; the standard susceptibility and dielectric constant measurements, on the other hand, must take account not only of the macroscopic interaction between the field and the magnetic or electric medium, but also of the interaction of the molecules with each other.

The spectroscopic method is that most closely allied to the method of molecular rays. The splitting of spectral lines in a magnetic field (Zeeman effect) or in an electric field (Stark effect) has, as is well known, yielded an enormous mass of data concerning the magnetic and electric properties of atoms and molecules. At the present stage of the technique, the deflection method is indeed, in the majority of cases, closely dependent on spectroscopic theory for an interpretation of its results. However, it actually surpasses the optical method in *sensitivity*; the reason is essentially that it is very much easier to produce high inhomogeneities than strong fields. Thus on the electrical side, the deflection method is often applicable

in cases where the corresponding Stark effect is too small to
be measurable; on the magnetic side, the use of a modified
form of the original Stern-Gerlach arrangement should make
possible the detection of magnetic moments which are ten or
even a hundred times smaller than those detectable in the
Zeeman effect with the most refined optical technique.

A new field has recently been opened up in the study of the
dissociation of metallic vapours by means of the magnetic
deflection method. If the atoms are magnetic, the molecules
non-magnetic, and a mixed molecular-atomic beam is shot
through an inhomogeneous magnetic field, the molecules go
straight on unaffected by the field; the atoms are deflected
right and left of the undeviated molecule beam. Actual
physical separation of the atoms and molecules can in this
way be effected, and their numbers can be counted by appropriate devices. The method has already yielded results of
great elegance (Chapter 7).

Other applications, which have thus far received but little
development, are also included in Chapter 7. We have summarised here only the main lines of advance, in order that a
general idea of the scope and importance of the molecular ray
method may be gained at the outset. But the technique is
advancing rapidly, and an application which is now obscure
may at any time assume quite another perspective. One may,
however, hazard the guess that in the next few years the
most important advances will come from a study of the wave
nature of molecular beams, both in its fundamental aspects
and in its application to the investigation of surfaces and the
mechanism of adsorption; while the possibility of working
with beams of sensibly uniform velocity should make possible
of attack many hitherto intractable problems, not least in
connection with the deflection method of studying the magnetic and electric properties of molecules.

Chapter 1

THE PRODUCTION AND MEASUREMENT OF MOLECULAR RAYS[a]

We have seen that an apparatus for the production of a molecular ray consists, schematically, of a vessel divided into three compartments by two diaphragms. The three compartments may be termed the source, the collimator chamber, and the observation chamber; the last named contains the detector.

The *source* may take any one of three forms, according to the nature of the substance under examination. (1) It may be a small oven, into which a limited quantity of the substance is introduced, to be vaporised after evacuation of the apparatus. (2) It may be simply a tube communicating through a capillary with a gas reservoir, from which the gas is drawn into the apparatus by pumps. (3) In certain special cases, it may be the surface of a molten solid or heated filament.

The *collimator chamber* is so called, in very loose analogy with optics, because its function is to form an approximately parallel beam of molecules. The technical problem in the design of the collimator chamber is the elimination of *alien molecules*, that is, those molecules which issue from the source aperture in directions other than the narrow range selected by the second or *image aperture*.

The design of the *observation chamber* naturally varies according to the specific purpose of the experiment. The common feature of all arrangements is the *detector*. The design of suit-

[a] The essence of this chapter is contained in three important papers from the Laboratory of Physical Chemistry at Hamburg: "Zur Methode der Molekularstrahlen, I", Stern, *Z. Physik.* **39**, 751, 1926 (U. z. M. 1); "Zur Methode der Molekularstrahlen, II", Knauer and Stern, *ibid.* **764** (U. z. M. 2); "Intensitätsmessungen an Molekularstrahlen von Gasen", Knauer and Stern, *ibid.* **53**, 766, 1929 (U. z. M. 10).

able detectors is to-day the central problem of the technique; much of the present chapter will be concerned with a description of the various forms of detectors which have so far been devised.

THE SOURCE

Molecular Effusion. The laws of molecular effusion, as distinct from hydrodynamic flow, of a gas through an opening in the wall of a containing vessel are very simply found; for the number of molecules effusing per second into an evacuated space through an aperture of area a is simply the number N of molecules which strike that area of the containing wall per second, namely

$$N = \tfrac{1}{4}n\bar{c}a \qquad \ldots\ldots(1\cdot1),$$

where n is the number of molecules per cubic centimetre and $\bar{c}$ is the mean velocity.

Two assumptions are inherent in equation (1·1). In the first place, the probability that a molecule arriving at the aperture passes through it has been set equal to unity; this implies that the thickness of the containing wall is negligibly small compared with the aperture dimensions. An aperture for which this condition is satisfied will be called an *ideal aperture*. Secondly, it is assumed that the uniform distribution of the molecules and of their velocity directions in the containing vessel are not disturbed by the act of effusion. The condition necessary for the stringent fulfilment of this second requirement is clearly that the aperture dimensions d shall be negligibly small in comparison with the mean free path λ of the molecules in the source; however, for most practical purposes the condition for molecular flow may be laid down as

$$\lambda \nless d \qquad \ldots\ldots(1\cdot2).$$

If the condition (1·2) is not fulfilled, there will be quasi-hydrodynamic flow at the aperture, and a turbulent gas jet, instead of a molecular ray, will be produced from such a source.

It is necessary to consider how far we are justified in applying (1·1) to the three chief types of source which are used in practice.

If the source is of the first, or oven type, it is essential to ensure that the charge, liquid or solid, has a sufficiently large surface. Clearly, if the surface is small compared with the area of the aperture, the rate of effusion is governed, not by (1·1), but entirely by the rate of evaporation of the charge. In practice, the available surface of the charge should be at least ten times the area of the aperture. The available surface is not necessarily identical with the geometrical surface of the charge; for example, unless special precautions are taken the surface of a liquefied easily oxidisable metal may be partially covered with an oxide film, through which evaporation is slow. It has in fact frequently been observed that on raising the temperature of an oven containing an insufficiently large solid charge to the melting point of the solid, there occurs a sudden drop in intensity, due to the very large decrease in the surface available for evaporation consequent on liquefaction.

In the second type of source, the gas might not attain temperature equilibrium with the walls if the rate of flow to the source aperture were too great; $\bar{c}$ in (1·1) would then cease to have a value corresponding to the temperature of the source. Actually, there is no such danger under practical conditions; for the flow velocity is perhaps 100 cm./sec., as against a molecular velocity of 10^4 to 10^5 cm./sec.

Turning to the third type of source, the surface of a molten solid or heated filament, it might appear at first sight that the average velocity of the molecules emitted from the surface would be less than that corresponding to the temperature of the source; for the molecules in escaping do work against the surface forces. Analysis of the equilibrium conditions for a substance in contact with its saturated vapour shows, however, that the velocity distribution of the molecules is unaffected by the act of evaporation: subject to the proviso that every vapour molecule on collision with the surface passes over into the other phase, an assumption legitimate so long as the surface is uncontaminated. The presence of surface forces has for its effect merely this: that only the fastest molecules escape, but in so doing are robbed of the greater part of their initial energy.[a]

The Cosine Law of Molecular Effusion. Equation (1·1) is explicitly concerned simply with the total quantity of gas effusing through an aperture per second; the present article deals with the angular distribution of the effusing molecules. In Fig. 2 let O be an ideal aperture, area a, piercing the dividing wall between a gas-filled vessel A and an evacuated space B, and suppose that the conditions for molecular flow through O are fulfilled. Then the number of molecules arriving per second at an element of area $d\sigma$ of a, within the solid angle $d\omega$ making an angle θ with the normal at $d\sigma$, which from elementary gas kinetic considerations is

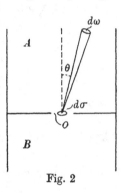

Fig. 2

$$\frac{n\bar{c}}{4\pi}.\cos\theta.d\omega.d\sigma,$$

is likewise the number effusing per second under the angle $d\omega$ from an ideal aperture. There is thus an exact analogy between the aperture regarded as a source of molecules and a source of radiation.

This, the cosine law of molecule effusion, is implicit in equation (1·1), the correctness of which was established experimentally by the pioneer work of Knudsen.[b] Recently the law has been verified directly by Mayer,[c] who explored the

[a] See Stern, Z. *Physik*, **3**, 418, 1920.

[b] Knudsen, *Ann. Physik*, **28**, 999, 1909.

[c] Mayer, Z. *Physik*, **52**, 235, 1928.

region in front of the aperture with a tiny radiometer-like vane suspended on a quartz fibre, the deflections of which are proportional to the number of molecules striking it.

INTENSITY IN THE BEAM

We proceed to calculate the *intensity* in a molecular beam, where by intensity is meant the number of molecules (or moles, according to the units chosen) passing through unit area of cross section of the beam per second.

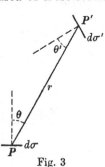

Fig. 3

The number of molecules which arrive at an element of area $d\sigma'$ at a point P', from the element of area $d\sigma$ at P ($PP' = r$, Fig. 3) of an aperture of area a is (Cosine Law of Radiation)

$$\frac{\nu/\pi . \cos \theta \cos \theta' . d\sigma \, d\sigma'}{r^2},$$

where $\nu = \frac{1}{4} n \bar{c}$; and the intensity directly opposite the aperture, at a distance r from it great compared with its dimensions, is thus

$$I = \frac{\nu a}{\pi r^2} \text{ molecules/cm.}^2 \text{ sec.} \qquad \ldots\ldots(1\cdot3).$$

This is likewise the intensity in the axis of a beam, at a distance r from the source.

It is often convenient to express the numerator of (1·3), which is merely the right-hand side of (1·1), as

$$N = \nu a = \frac{N_0}{\sqrt{2\pi RMT}} . pa \text{ molecules/sec.} \quad \ldots(1\cdot4),$$

where N is the number of molecules effusing per second, N_0 is Avogadro's number, R the gas constant per mole, M the molecular weight of the gas, T the absolute temperature of the source, and p the pressure in the source in dynes/cm.²

The number q of moles effusing per second is

$$q = \frac{1}{\sqrt{2\pi RMT}} . pa \text{ moles/sec.} \qquad \ldots\ldots(1\cdot5 \, a),$$

if p is expressed in dynes/cm.^{2}; or

$$q = \frac{1}{\sqrt{2\pi RMT}} . 1\cdot33 \times 10^3 . pa$$

$$= \frac{5\cdot83 \times 10^{-2}}{\sqrt{MT}} . pa \text{ moles/sec.} \quad \ldots\ldots(1\cdot5\ b),$$

if p is expressed in millimetres of mercury. $(1\cdot5\ b)$ is the form most frequently used in the sequel.

Slits. It is now clear, from $(1\cdot3)$ and $(1\cdot4)$ or $(1\cdot5)$, that $I \propto p.a$. Thus to meet a demand for high intensity, it is necessary to make the product $p.a$ as large as possible. Now in many applications of the molecular ray method, most obviously in the measurement of magnetic or electric moments (Chapters 5 and 6), the sensitivity of the method is increased by making the beam narrow; for the narrower the beam the smaller will be the deflections detectable by experiment. This demand and that for maximum intensity can be simultaneously met by giving the source aperture (and also the image aperture) the form of a slit.

An upper limit to p is set by the condition $(1\cdot2)$. In the case of a slit aperture, the quantity d may be taken as the slit width b; for it is found in practice that collisions in the direction of the slit length are relatively unimportant in their effect on the definition of the beam. Here emerges the most distinctive and valuable feature of the slit aperture; namely, that the beam intensity is independent of the slit width. Thus suppose the slit width is halved: then it is true that the area a is also halved, but at the same time it is now permissible to double the pressure; that is, the product $p.a$ and therewith I remains unaltered. Contrast with this the consequences of halving the diameter of a circular aperture: a becomes $a/4$, but in compensation it is possible only to double the pressure, and the product $p.a$ is halved.

So far, an ideal slit has been tacitly assumed. The slits used in actual practise, however, must necessarily depart to a greater or less extent from the ideal; that is, they will have in general the form of a short canal. The quantity of gas effusing molecularly at a given pressure through short slit-shaped canals has been calculated by Clausing.[a] Table I is taken from his paper. Infinitely long slits are assumed, and the quantity W, namely the probability of a molecule arriving at the aperture passing through it, is tabulated against the ratio of the length of the canal L to the width b.

Table I

L/b	W	L/b	W	L/b	W
0	1	1·3	0·6321	3·2	0·4439
0·1	0·9525	1·4	0·6168	3·4	0·4318
0·2	0·9096	1·5	0·6024	3·6	0·4205
0·3	0·8710	1·6	0·5888	3·8	0·4099
0·4	0·8362	1·7	0·5760	4	0·3999
0·5	0·8048	1·8	0·5640	5	0·3582
0·6	0·7763	1·9	0·5525	6	0·3260
0·7	0·7503	2·0	0·5417	7	0·3001
0·8	0·7266	2·2	0·5215	8	0·2789
0·9	0·7049	2·4	0·5032	9	0·2610
1·0	0·6848	2·6	0·4865	10	0·2457
1·1	0·6660	2·8	0·4712		
1·2	0·6485	3·0	0·4570	∞	$\dfrac{b}{L} ln \dfrac{L}{b}$

It appears therefore at first sight that, far from being independent of the slit width, the intensity is in practise seriously diminished by narrowing the slits. This is fortunately not necessarily the case. It is true that the *total quantity* of gas effusing per second is diminished; against this, however, must be set the fact that with a canal-like aperture the *directions* of the effusing molecules are clustered more closely than is the case for a cosine law of effusion about the direction of the axis of the canal. The angular distribution of the molecules effusing through a short *circular* canal ($L < r$, or of the same order as r; $r =$ radius) has been calculated by Clausing.[b]

[a] Clausing, *Physica*, 9, 65, 1929.
[b] Clausing, Z. *Physik*, 66, 471, 1930.

His results for the case $2r = L$ are exhibited graphically in Fig. 3'. The full curve shows the actual distribution; the dotted curve that satisfying the cosine law. The pointed formation of the full curve is worthy of note. It will be observed that the area under the full curve is only 0·512 times that under the dotted curve.

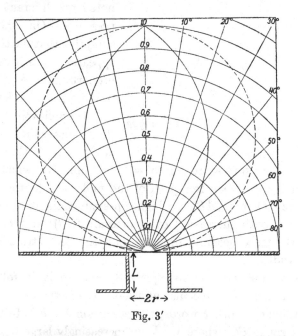

Fig. 3'

Mayer,[a] using his torsion balance method, has observed experimentally a departure from the cosine law of effusion with canal-like apertures, in the direction of an increased concentration of the effusing molecules about the direction of the axis of the canal. His results are in qualitative agreement only with the theory.

Cloud Formation. It remains to discuss rather more fully the limiting condition (1·2) with reference to the attainable beam intensity. The question has been approached experi-

[a] Mayer, Z. Physik, 58, 373, 1929.

mentally by Knauer and Stern.[a] Water was evaporated from an oven having a slit 6 mm. long and 0·022 mm. wide; an image slit of similar dimensions defined a beam which was received on a liquid air-cooled silver target as a streak-like deposit. The time of appearance of the trace was noted for different oven pressures. On the assumption that the time of appearance t is inversely proportional to I, the product $p.t$ is, so long as (1·5) is applicable, a constant. The results graphed in Fig. 4 show that such is the case up to a pressure of some 0·5 mm. Hg, which corresponds to a mean free path of the molecules in the oven of some 0·06 mm. (AB). At higher pressures, the product $p.t$ increases with increasing oven pressure (BC); at the same time the trace becomes broader than the geometrically defined width.

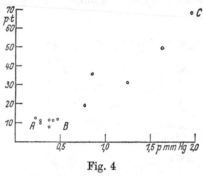

Fig. 4

Knauer and Stern have explained these results as follows: In the range AB, molecular effusion through the oven slit is maintained; in other words, the condition $\lambda \nless b$ is fulfilled. In the range BC, where $\lambda < b$, an increasingly large fraction of the effusing molecules collide with one another in the slit itself or immediately in front of it, whereby they are dammed up outside the oven slit, forming there a cloud of vapour (Wolke). The effective boundary of the cloud, and not the slit, is then the radiating source. The radiating surface of the cloud is naturally greater than the surface of the oven slit; moreover, the number of molecules effusing must be smaller than if no cloud were there: so that the "surface brightness" of the cloud is *a fortiori* less than that of the oven slit.

a Knauer and Stern, *Z. Physik*, **39**, 774, 1926 (U. z. M. 2).

Result, a falling off in the intensity and a broadening of
the beam.

Mayer[a] has called this interpretation in question. He examined by his torsion balance method the distribution with
angle of the molecules effusing from a slit (0.01×0.05 mm.)
closely approaching the ideal, at an average distance of about
5 mm. from the slit, and found the cosine law to hold up to
oven pressures of about 4 mm. He infers that the vapour
composing the cloud postulated by Knauer and Stern is
literary rather than scientific. This conclusion is not justifiable,
as has been pointed out by Knauer and Stern,[b] and that on
two scores: (1) The length of their slits was often 400 times
the width; in such case it is easy to show that the intensity
in the immediate neighbourhood of the slit falls off as $1/r$;
with Mayer's short slits, $I \propto 1/r^2$ approximately. Thus with
$\lambda \cong \frac{1}{4}b$, as in his experiments, cloud formation must have
been at the most very slight. (2) Even assuming the existence
of a cloud, Mayer working as he did *with no image slit* would
measure the "total brightness" of the cloud, not the "surface
brightness" as in the molecular ray experiments. Thus the
possibility of finding a cosine law at an observing distance of
5 mm. is even with cloud not excluded.

Johnson[c] has also criticised Knauer and Stern, but the experimental results which he brings in support of his arguments
are in sharp contradiction to the experience of other workers:
namely, that with an oven slit 1 mm. by 0.1 mm., the intensity of a mercury beam increased *linearly* with the oven
pressure p up to $p \cong 30$ mm.

It is clearly desirable that this question should be settled
conclusively. In this connection, direct evidence of cloud
formation might be obtained by illuminating the cone of rays
issuing from an oven containing sodium or mercury with the

a Mayer, *Z. Physik*, **58**, 373, 1929.
b Knauer and Stern, *Z. Physik*, **60**, 414, 1930.
c Johnson, *Phys. Rev.* **31**, 103, 1928.

appropriate resonance lines. The resonance radiation would show up the form of the cone at the lowest pressures; the shadow of the cone could be utilised for the same purpose at the higher pressures.

THE COLLIMATOR CHAMBER

The molecules issue from the source slit as a conical pencil of rays, whose maximum intensity is in the direction normal to the slit (cf. Fig. 3′); the function of the image slit is to select from the total those molecules whose directions coincide, within limits determined by the geometry of the apparatus, with that of the normal.

Now just as scattered light must be eliminated from an optical system, so it is desirable, if one would secure maximum definition of a molecular beam, to effect the removal of alien molecules from the collimator chamber. Their presence leads moreover to a further undesirable consequence, for which there is no optical analogy: reflected from the walls, they set up a gas pressure, which not merely impairs definition, but actually weakens the beam by deflecting molecules which would otherwise travel on through the image slit out of their proper path.

Elimination of Alien Molecules

There are two ways of removing the alien molecules from the collimator chamber: either (1) by cooling the walls to a temperature low enough to condense them on the initial impact, or (2) by rapid pumping. In most laboratories, the first method is limited in its application to substances having a sufficiently small saturation pressure (say 10^{-6} mm.) at the temperature of liquid air; in some cases (for example, hydrogen or helium) it is of course not possible to use it at all. Where a choice lies between the two methods, the first is greatly to be preferred, as will be clear from a numerical example: A normal apparatus easily permits the introduction

of 100 cm.[2] of cooled surface between the source and image
slits; whereas a pump having the very high effective speed
at the collimator chamber of 10 litres/sec. at a pressure of
10^{-3} mm. is equivalent on the average to a perfect condenser
of rather less than 1 cm.[2] Thus the first method may easily
be a hundred times as efficient as the second.

First Method. The collimator chamber must be designed to
accommodate a sufficiently large condenser. On the assumption that every molecule is condensed on its first impact, it is
easy to show that the partial pressures in source and collimator chamber are in inverse ratio to the areas of source
slit and condenser. Thus if a minimal pressure of p' mm. in
the collimator chamber is aimed at, the condenser must have
an area of $p.a/p'$ cm.[2] Numerical example: $p = 0.1$ mm.,
$a = 10^{-2}$ cm.[2], $p' = 10^{-5}$ mm. Area necessary, 100 cm.[2]

The temperature at which the condenser must be held is
decided by the vapour pressure curve of each particular substance. Thus for example Hg, H_2O require liquid air temperature, Na, K, Cd need water cooling only, while Ag, Cu,
Bi condense readily enough at room temperature.

Second Method. Here a gas to which the condensation
method is inapplicable is presupposed; thus the source is of
the second type, where the gas flows from a reservoir via a
capillary to the source slit. For a given rate of flow and pump
speed there will be a certain stationary pressure $p = p_s$ in the
collimator chamber. If I is the intensity of the beam at the
image slit, distant l from the source, when $p = p_s$; and if I_0
is the value which the intensity would have if there were no
weakening of the beam, then

$$I = I_0 e^{-(l/\lambda_{p_s})} \qquad \ldots\ldots(1.6),$$

where λ_{p_s} is the mean free path when $p = p_s$. Now clearly
$I_0 \propto p_s$; hence I is a maximum ($I_{max} \cong \frac{1}{3}I_0$) for $\lambda_{p_s} = l$ (cf.
Chapter 2, p. 76), a condition which serves to determine
the optimum p_s. Thus for a given distance l between source

and image slits, the rate of flow to the source slit and there-with the intensity attainable is proportional to the pump speed.

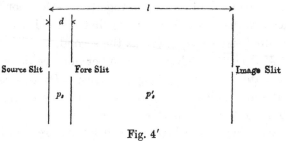

Fig. 4′

Geometrical considerations set a practical minimum to l (see p. 51). It is nevertheless possible greatly to increase the available intensity by introducing a third slit (or *fore slit*) at a short distance d from the source slit, thus dividing the collimator chamber into two parts, the antechamber and the collimator chamber proper (Fig. 4′). If the two chambers are connected to separate pumps, the stationary pressure in the antechamber will remain equal to p_s, but that in the collimator chamber proper will have the much smaller value p_s'. In actual practice, p_s is of the order 10^{-3} mm., p_s' of the order 10^{-4} mm.

It appears at first sight that it should be possible after introduction of the fore slit to increase the rate of flow and hence the intensity l/d times. This is however not the case. Thus Knauer and Stern,[a] with an apparatus in which l was 4 cm., observed a fourfold increase in intensity of introducing a fore slit at a distance of 4 mm. from the source slit; not as expected a tenfold increase ($l/d = 10$). The loss in intensity evidently arises from alien molecules reflected from the partition wall in the immediate neighbourhood of the fore slit. These molecules are particularly troublesome because, being reflected preponderantly at right angles to the wall (see

[a] Knauer and Stern, *Z. Phys.* 53, 777, 1929 (U. z. M. 10).

Chapter 3), their direction of motion is opposite to that of the beam molecules. Hence the chance that they make effective scattering collisions with the latter is unusually great, and can clearly become important when d is small.

An approximate theory can be constructed somewhat as follows. At a given temperature and pressure, $\lambda \propto 1/\nu = \dfrac{1}{c.\nu}$, where ν is the number of molecular collisions per cm.³ per sec. We can therefore rewrite (1·6) as

$$I = I_0 e^{-d.c\nu} \qquad \ldots\ldots(1\cdot7),$$

where ν refers to collisions suffered by the beam molecules over the distance d. ν is made up of two groups: a number $\nu_1 \propto p_s$ of collisions with alien molecules in de-coordinated motion in the antechamber, and a number ν_2 of collisions with alien molecules directly reflected from the partition wall. Now $\nu_1 \propto N/S = c_1 N/S$, where N is the number of molecules effusing per second from the source, and S is the pump speed. ν_2 is proportional to the number of molecules which in unit time strike unit area of the dividing wall in the immediate neighbourhood of the fore slit, that is $\nu_2 = c_2 . N/\pi d^2$. Inserting these values in (1·7), we have

$$I = I_0 e^{-cd(\nu_1+\nu_2)} = I_0 e^{-cdN(c_1/S + c_2/\pi d^2)} \qquad \ldots\ldots(1\cdot8).$$

Now $I_0 \propto N = c_3 N$; hence I is a maximum for a given value of d when

$$N = \frac{1}{cd\,(c_1/S + c_2/\pi d^2)}$$

and $\qquad I_{\max} = \dfrac{c_3}{cd\,(c_1/S + c_2/\pi d^2)} . e^{-1} \qquad \ldots\ldots(1\cdot9).$

The optimum value of d is that for which the denominator in the right-hand side of (1·9) is a minimum, namely

$$d = \sqrt{\frac{c_2}{\pi c_1} . S};$$

when $\nu_1 = \nu_2$. Thus finally

$$I_{\text{optimum}} = C\sqrt{S} . e^{-1} \qquad \ldots\ldots(1\cdot10),$$

where $C = 2\sqrt{c_1 c_2/\pi} . cc_3$.

It emerges from the theory first, that for a given pump speed there is an optimum value of d, when the effects of the random and reflected molecules in weakening the beam are just equal. Further, there is the important result that at optimum d the intensity attainable is proportional to the square root of the pump speed. It remains of course true that for large values l of d, when the directly reflected molecules play a negligible part in weakening the beam, the intensity attainable is directly proportional to the pump speed. This can be seen immediately from equation (1·9), if the second term under the bracket in the denominator is neglected.

The fundamental importance of the pump speed for the efficiency of the second method makes it essential to use the fastest available pumps, and to get the most out of them by using wide and short connecting tubes. Even then, the maximum intensity attainable is in the meantime still determined by the rate at which the pump can remove the alien molecules from the collimator chamber, and not by the natural condition imposed by (1·2).

DETECTORS

The minimum beam intensity which can be detected is of fundamental importance for the sensitivity of the molecular ray method. That this is so in general is of course evident without special argument from the fact that a physical limit to the intensity attainable is set by the source condition $\lambda \nless d$; but a more vivid impression comes from a consideration of particular cases, from which we select with Stern the measurement of atomic magnetic moments by the deflection method (Chapter 5) as typical.

A beam of atoms, magnetic moment μ, is shot through an inhomogeneous magnetic field. An atom in the beam is, while in the field $\left(\text{inhomogeneity } \dfrac{\partial H}{\partial s}\right)$, acted on by a force

$$\mathbf{F} = |\,\mu\,| \, \frac{\partial H}{\partial s}$$

perpendicular to the beam direction. This force imparts to an atom of mass m an acceleration of amount $f = F/m$, and the beam suffers a deflection

$$s = \frac{1}{2} \cdot ft^2 = \frac{1}{2} \cdot \frac{F}{m} \cdot \frac{l^2}{v^2},$$

where l is the length of path in the field, and v is the velocity of the atoms.

Hence the sensitivity attainable (that is the minimum μ detectable) depends on:

1. A narrow beam; for the narrower the beam, the smaller is the measurable deflection.

2. A long flight in the field; for $s \propto l^2$.

3. A large deflecting force, for $s \propto F$.

We have seen that it is possible to have a narrow beam, and still retain intensity, by using long narrow slits. From equations (1·3) and (1·5)

$$I = \frac{q}{\pi r^2} = \text{const.} \frac{p \cdot a}{r^2} = \text{const.} \; pb \cdot \frac{h}{r^2},$$

where h is the length of the slit. If we set b equal to 0·1 mm., then the pressure p corresponding to a mean free path of 0·1 mm. is roughly 0·1 mm.; thus at the optimum pressure $p \cdot b = 10^{-3}$, and

$$I \propto \frac{h}{r^2}.$$

Next, let us examine conditions 2 and 3 above. First, $s \propto r^2$. Second, $s \propto F \propto \dfrac{\partial H}{\partial s}$; we must therefore make the inhomogeneity of the field as great as possible. This we can do by aiming the beam a very short distance a from a wedge-shaped pole piece, for in such case $\dfrac{\partial H}{\partial s} \propto \dfrac{1}{a}$. On the other hand, this high degree of inhomogeneity only exists in a region whose cross section is of the order a^2; therefore an upper limit $h = a$ is set to the slit length. Hence the force $F \propto 1/h$, and we have finally

$$s \propto \frac{r^2}{h}.$$

Thus $s_{max} \propto 1/I_{min}$, a particular condition which emphasises the extreme importance for the technique of the minimum intensity detectable.

This fundamental importance of the minimum detectable intensity to the sensitivity of the molecular ray method makes the design of sensitive detectors the central problem on the technical side. The quest is moreover towards detectors which are not merely sensitive, but metrical. It is fortunately not of immediate importance to make absolute measurements of the intensity. Provided that the intensity of the parent beam is held constant, it is possible to refer the intensities of molecules deflected, scattered, or reflected in particular directions to the parent intensity; one is thus involved merely in relative measurements, which are of course incomparably the easier to make.

Detectors fall, broadly speaking, into two classes: the additive class, to which the optical analogy is the photographic plate; and the non-additive class, paralleled in optics by the photo-electric cell or the thermopile. Detectors typical of both classes may on occasion be used to supplement each other in a single investigation; an additive type often yielding a valuable general survey of a situation which is to be subjected to detailed quantitative study by a non-additive method.

ADDITIVE DETECTORS

The Condensation Target

A glass or metal plate is placed in the path of the beam, and maintained at a temperature low enough to allow the impinging molecules to accumulate on its surface as a permanent deposit. This condition fulfilled, it is only a question of time before the deposit becomes visible. The length of exposure is however limited, except for the second type of source, by the amount of substance in the source, and may not in all cases be sufficient to give a visible deposit. In that case, the invisible deposit can often be "developed" by appropriate methods.

Transition Intensity. The temperature at which a particular target must be held depends in the first instance on the substance under examination; thus for example mercury requires under comparable conditions a much lower target temperature than silver. For any given substance, it depends primarily (*a*) on the beam intensity, (*b*) for a given beam intensity, on the width of the beam.

(*a*) Given target material and temperature, there exists for each beam species a *transition intensity*,[a] below which a permanent (that is, a developable) deposit is not formed no matter how long the target is exposed to the beam (see Chapter 3, p. 93). Thus a target temperature perfectly suitable for the detection of the parent beam may be too high for the detection of such small fractions of it as are studied for example in the scattering of molecular rays from solids.

(*b*) The molecules after striking the target do not remain in one fixed position, even if the temperature of the target is well below the transition temperature necessary for the formation of a permanent deposit; they move about at random on the surface (see Chapter 3, p. 90). At the centre of the deposit, as many molecules move into a given small area as leave it, and the nett effect of the surface motion is nil. It is otherwise at the edges; there the drift of molecules across the border is predominantly in the direction of emigration. With a very narrow streak-like deposit this lateral motion (Rutscheffekt)

[a] I have used the terms "transition intensity" and "transition temperature" in preference to the customary "critical intensity", "critical temperature", reserving the term "critical" to refer to the surface two-dimensional gas (see Chapter 3).

may become important enough to affect the transition intensity; and indeed the transition intensity is actually found to be higher for the narrower beams (see Chapter 3, p. 91). Thus for a given I_{min}, the condensation target detector sets a physical (as distinct from a technical) limit to the slit width b; we have $b_{min} \propto I_{min}$.

Target Material. Our knowledge of how the target material affects the transition intensity is entirely empirical, and owing to the difficulty of preparing uncontaminated surfaces, not entirely reliable. There are some obscure hints that chemical effects may be important: thus sodium adheres strongly to glass, mercury to silver, etc. The question is not however alone that of choosing a substance for which I_{min} is the smallest possible: the visibility of the deposit must be good. For example, a substance deposits very readily on itself,[a] but this property, while being useful as a means of *sensitising* a surface, is obviously no guide to the choice of a target material. Again, sodium deposits more readily on glass than on (e.g.) silver, but the relatively poor visibility of a sodium-on-glass deposit decides against a glass target. In practice therefore one constructs a test target made up of narrow contiguous strips of different metals, and observes the order of appearance on the various metals of a trace running perpendicular to the strips. Thus on a composite target of gold, copper, nickel, silver a visible deposit from a bismuth beam appeared first on the nickel, and then in the order copper, silver, gold.

The range of application of the condensation target is in most laboratories limited to substances for which the beam intensity under the conditions of experiment is above the transition intensity for a target held at liquid air temperature.

Developers. The target conditions for the formation of a permanent deposit complied with, it remains to ensure the growth of a clearly *visible* deposit, most simply by providing

[a] See for example Donat and Philipp, *Z. Physik*, **59**, 6, 1930.

an adequate length of exposure. Where this is not feasible, recourse may be had to either of two methods of developing the invisible trace.

METHOD 1

A "wet" method. Its use is restricted to metallic deposits, and the target must be of glass. The target with its invisible trace is removed from the apparatus and placed in a bath of not too fresh 1 to 2 per cent. hydroquinone solution containing gum arabic as protective colloid. Addition of a few drops of a 1 per cent. silver nitrate solution causes separation of atomic silver, preferentially at places where a metallic deposit is already present. Except it is very weak, the trace bursts into view after a few minutes in the developer. Invisible deposits of the more noble metals, such as Cu, Ag, Au, Ni retain their geometrical form on development; deposits of hygroscopic or easily oxidisable metals, as Fe, Pb, Tl, cannot be removed from the apparatus for development without suffering serious distortion.[a]

METHOD 2

The target, which may be metallic, remains *in situ* in the apparatus. A metallic vapour at low pressure is admitted to the apparatus (in practice, removal of the liquid air from a mercury trap serves very well), when the vapour condenses preferentially on the nuclei furnished by the deposit. Result, a faintly visible trace is intensified, an invisible trace is rendered visible. Unless over developed, the deposits of all substances so far studied retain their geometrical dimensions. The method is undoubtedly capable of considerable refinement, and might conceivably be made quantitative. Even in a crude form it is preferable to Method 1.

The method may also be used to *sensitise* the target. If mercury vapour is allowed access for a time to the cooled

[a] Estermann and Stern, Z. *physikal. Chem.* **106**, 399, 1923; Gerlach, *Ann. Physik*, **76**, 179, 1925.

target *before* its exposure to the beam, the time required for the appearance of a visible trace is considerably shortened.[a]

Appearance of the Traces. The appearance of the trace is practically independent of the substance. When just visible it is light brown, on further exposure it becomes gradually darker till black, and finally on prolonged exposure bluish with a metallic sheen. If fairly strongly developed (Method 2) it may appear light on a dark background (*i.e.* negative); on further development the change positive-negative may occur several times. Visibility of the trace is often particularly good in the second negative stage.

Plate I shows typical photographs of traces. Fig. 5 shows a number of mercury traces on copper; trace 4 and its right-hand neighbour show the effect of too high an oven pressure (turbulence in the beam). Fig. 6 shows traces of a lithium beam split into two by a magnetic field; target material, nickel. Fig. 6 *a* shows the double trace shortly after its appearance. Fig. 6 *b* shows the same trace after a long exposure; the umbra can be clearly recognised, metallic against the blue-black background of the penumbra.

Sensitivity. It has been found that in a deposit from a molecular beam the molecules are not distributed uniformly over the area of the deposit, but are concentrated in isolated condensation nuclei each consisting of many thousands of molecules (Chapter 3, p. 90). This peculiarity of the structure of the deposit makes the condensation target an extremely sensitive detector; in favourable cases a deposit with a *calculated* thickness of two molecules is, thanks to the molecular aggregation on the surface, actually visible. In the sequel, the expression "a deposit n molecules thick" refers always to the thickness calculated on the assumption of a uniform distribution of molecules over the surface.

Let us calculate the time of appearance of the trace, on the

---

[a] Estermann and Stern, *loc. cit.*; Knauer and Stern, *Z. Physik*, **39**, 778, 1926 (U. z. M. 2).

Plate I

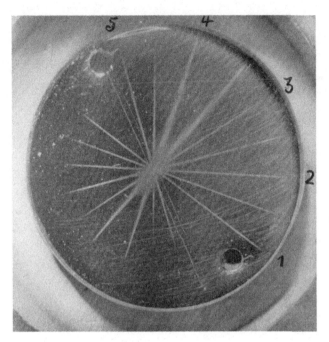

Fig. 5

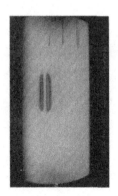

a b

Fig. 6

Typical Traces

assumption that a deposit two atoms thick is visible, for the case of silver. From (1·3) and (1·4), we have, setting $p \cdot b = 10^{-3}$ (p in mms.),

$$I = \frac{1 \cdot 1 \times 10^{19}}{\sqrt{MT}} \cdot \frac{h}{r^2} \text{ molecules/cm.}^2 \text{ sec.}$$

Thus with $M = 108$, $T = 1300°\ K$, $h = 1$ cm., $r = 10$ cm., we have $I_{Ag} = 3 \times 10^{14}$ molecules/cm.^{2} sec. A monomolecular layer contains some 10^{15} molecules per cm.^{2} Therefore, on the assumption that the target temperature is so low that every molecule condenses on impact, a deposit two molecules thick will have formed in $\dfrac{2 \times 10^{15}}{3 \times 10^{14}} \cong 10$ sec.

The wet method of development increases the above sensitivity about 10 times; the dry method under the most favourable conditions, the realisation of which has so far not been attempted in molecular ray work, as much as a thousand times.[a]

Possibility of Quantitative Measurements. The formation of thin deposits is, as we shall see in Chapter 3, such an extremely complicated and little understood process that the condensation target, in spite of its very great sensitivity, must remain for the present at most a semi-quantitative detector: and that for the following reasons:

(1) In the above estimation of the optimum sensitivity, it was assumed as a basis for calculation that every molecule condenses on impact. In actual fact, unless the beam intensity is several times the transition intensity, a fraction of the incident molecules dependent on constants characteristic of the beam species re-evaporate.[b] Thus in attempting to estimate the intensity from the thickness of the deposit these constants must first be known.

[a] Langmuir, *Proc. Nat. Acad. Sci.* **3**, 141, 1914.

[b] Cockcroft, *Proc. Roy. Soc.* **119**, 303, 1928. Cockcroft's data refer strictly to the deposition of cadmium on copper only, but qualitatively the results are probably quite general.

(2) The characteristic curve connecting deposit thickness with number of molecules present has never been accurately determined; it is not enough to assume it to be linear.

(3) The speed of lateral motion is unknown, and hence its effect on the narrower beams cannot be taken quantitatively into account.

(4) It is extremely difficult to maintain a reproducible target surface, or even one which has steady characteristics over the time of a long exposure.

Four chief methods have been used in the attempt to determine beam intensities with the target detector. In assessing their usefulness, the difficulties enumerated above must be taken into account.

(1) Where it is possible to collect a relatively thick deposit (over say 200 Å.) the thickness may be determined interferometrically by the Wiener method correct to 10–20 Å. A metal must first be converted to one of its salts, to ensure a definite change of phase $\pi/2$ at the surface (see Table II).

(2) In particular cases, direct weighing of a non-oxidisable, non-hygroscopic deposit on a microbalance is sometimes useful. Expressing I_{Ag} of the numerical example above in grams per cm.^{2} per second, $I_{Ag} = 5 \times 10^{-8}$ gm. Ag/cm.^{2} sec. For a beam of cross section 10^{-3} cm.^{2}, 3×10^{-8} gm. collect in 10 minutes; a very sensitive microbalance responds to about 4×10^{-9} gm. It is clear that the method, even for a heavy metal like silver, is relatively insensitive. However, with wide beams and long exposures it offers no special difficulty and has been used with some success in at least two instances (see Table II).

(3) Direct photometry of the deposit has been used, rather lightheartedly, by some workers. It is clearly a method which must be approached with extreme caution.

(4) The most generally useful procedure is to use the *time of appearance t* of the trace as a measure of the beam intensity: $I \propto 1/t$. In view of what has been said, this is of

course only very approximately true. Moreover, quite apart from physical processes, t obviously depends on a number of physiological factors, such as the ability to distinguish minute differences in surface brightness, eye fatigue, and so on.

Chemical Targets

Here the impinging molecules react chemically with the substance composing the target. The reaction product must give a good colour contrast with the unaffected surface. Typical of this method are (1) the reception of atomic hydrogen on a MoO_3 target, giving a blue MoO_2 trace against the pale yellow MoO_3 background; (2) the reception of atomic oxygen on litharge: contrast, brown PbO_2 against pale yellow PbO. The sensitivity of both these targets is quite good, the trace appearing in a few minutes with the target at some 15 cm. from the source.

It is probable that there exists for the chemical target a critical beam intensity below which a visible trace of the beam does not result even on indefinitely prolonged exposure. Thus for example the magnetically deflected traces of atomic hydrogen cease widening after a certain length of exposure.

An attempt has been made by Johnson[a] to develop the MoO_3 method of detecting atomic hydrogen metrically. He finds that the darkening, as measured on a densitometer, is directly proportional, to within $\pm$ 25 per cent., to the number of hydrogen atoms striking the target, provided that over-exposure is avoided.[b]

Kerschbaum[c] claims to have found in the Schumann plate an extremely sensitive detector for atomic hydrogen. He states that under similar test conditions a MoO_3 target began to go blue in 3 to 4 minutes, whereas a Schumann plate was completely blackened in $\frac{1}{2}$ to 1 minute. These figures are to be accepted with considerable reserve.

[a] Johnson, *J. Franklin Inst.* **207**, 629, 1929.

[b] *Ibid.* **210**, 141, 1930.

[c] Kerschbaum, *Ann. Physik*, v, **2**, 201, 1929.

NON-ADDITIVE DETECTORS

Manometers [a]

For gases such as H_2, He, O_2, etc., the beam may be received at a narrow slit, the only opening to an otherwise closed vessel. Entry of the beam sets up a pressure in the vessel, which is measured by a suitable manometer. The maximum stationary pressure p_∞ is determined by the condition that as much gas leaves the slit per unit time as the beam brings in. The quantity of gas entering the slit is, by equations (1·3) and (1·5),

$$\text{const. } \frac{pa}{\pi r^2} \cdot \frac{1}{\sqrt{T}} \cdot a',$$

where a' is the area of the receiving slit; while the quantity leaving the vessel at the temperature T' is

$$\text{const. } \frac{p_\infty a'}{\sqrt{T'}}.$$

Thus $\qquad\qquad p_\infty = \frac{p \cdot a}{\pi r^2} \cdot \sqrt{\frac{T'}{T}} \qquad$ ......(1·11′).

Numerical example: $p = 1$ mm., $a = 10^{-3}$ cm.^{2}, $r = 10$ cm.; then if $T = T'$, $p = 3·18 \times 10^{-6}$ mm., a quite practicable pressure.

κ FACTOR

A considerable increase in sensitivity is gained by giving the receiving or *detector slit* the form of a canal. This has no effect on the entering beam, since it is undirectional; but a greater resistance is offered to the exit of the de-coordinated molecules from the vessel. p_∞ is therefore increased by a factor $\kappa = \frac{R_c + R_s}{R_s}$, where R_c, R_s are respectively the Knudsen resistances of the canal, and of an ideal slit of equal cross section. In practice, κ may be as large as 10, so that we have to deal now with a stationary pressure of some 10^{-5} mm.

[a] See Knauer and Stern, Z. *Physik*, **53**, 766, 1929 (U. z. M. 10).

PRESSURE FLUCTUATIONS

Suppose we aim at measuring the parent beam correct to 1 in 1000; this is equivalent to measuring pressure differences of 10^{-8} mm. in a total pressure of 10^{-5} mm. Now in using pumps to eliminate the alien molecules, the remanent pressure in the observation chamber is perhaps 10^{-5} to 10^{-6} mm., subject to random variations arising from fluctuations in the pump speed of as much as 10 per cent. The variations are therefore of the same order or greater than the pressures we want to measure. Trouble from this source can be decreased considerably by inserting a large buffer vessel (ca. 12 litres) between pump and observation chamber. It can be practically entirely eliminated by taking the further step of attaching to the apparatus a compensation manometer, as far as possible identical with the measuring manometer, communicating with the observation chamber through an adjustable Knudsen resistance. This may take the form of a tap having a shallow channel cut in the barrel, extending from the bore about a quarter way round the circumference. The tap is turned until the Knudsen resistance of the channel is equal to that of the detector slit. When this is the case, a sudden fluctuation of pressure in the observation chamber travels at equal rates to the two manometers, and the zero of the arrangement is unaffected.[a]

ZERO CREEP

A steady zero creep can be eliminated in the standard way by taking timed readings with the beam alternately free to enter the measuring manometer and cut off by a shutter. This imposes an essential condition on the manometer: the final pressure must be rapidly attained.

[a] Professor Stern informs me that still better results are obtained if the compensation manometer is built initially as the exact counterpart of the measuring manometer. The adjustable tap may then be dispensed with.

TIME LAG

The time lag of the manometer is very simply calculated. If n is the instantaneous number of moles in the volume V of the manometer, the instantaneous pressure $p_i = n \cdot RT'/V$; and the increase of pressure per second

$$\frac{dp_i}{dt} = \frac{RT'}{V} \cdot \frac{dn}{dt}.$$

Now dn/dt is equal to the number of moles $I_r a'$ entering the manometer per second, less the number leaving per second

$$\frac{p_i \cdot a'}{\kappa \sqrt{2\pi RMT'}}.$$

Thus $$\frac{dp_i}{dt} = \frac{RT'}{V} \cdot a' \left(I_r - \frac{p_i}{\kappa \sqrt{2\pi RMT'}} \right).$$

For $p_i = p_\infty$, $dp_i/dt = 0$; that is,

$$p_\infty = I_r \cdot \kappa \sqrt{2\pi RMT'} = \kappa \cdot pa/\pi r^2$$

when $T = T'$, by (1·5). This is in agreement with the value found directly in (1·11') above. Therefore

$$\frac{dp_i}{dt} = \frac{a'}{\kappa V} \cdot \sqrt{\frac{RT'}{2\pi M}} \cdot (p_\infty - p_i);$$

whence the time required for the pressure in the manometer to attain the value p_i is

$$t = \frac{\kappa V}{a'} \cdot \sqrt{\frac{2\pi M}{RT'}} \cdot 2\cdot303 \log_{10} \frac{p_\infty}{p_\infty - p_i} \quad \ldots(1\cdot11).$$

Thus whereas p_∞ is independent of the area a' of the detector slit, $t \propto 1/a'$. In magnetic or electric deflection experiments the length h of the slit is limited in practice to about 2 mm. (cf. p. 25 above); this sets an upper limit to a', and with the heavier gases ($t \propto \sqrt{M}$) the time lag can become wearisomely great. p_∞ and t are both proportional to κ; it is therefore for practice to decide how large κ can profitably be made.

Numerical example: $a' = 1 \times 10^{-3}$ cm.^{2}; effective $V = 35$ c.c. (see p. 40 below); $\kappa = 10$; $M = 2$ (Hydrogen); $T' = 300°$ K.: then the time required for the pressure to attain 99 per cent. of its final value is 35 seconds.

The above calculation assumes (1) that the inherent lag of the manometer (that is, its time of response to an *instantaneous* change of pressure) is negligible (see further p. 39); (2) that the gas is not adsorbed on the walls of the manometer vessel. The effect of adsorption is clearly to increase the effective volume V of the vessel, and with it the time lag. The latter may indeed become so great as to make the manometer method impracticable unless means can be devised of reducing the adsorption.

CALIBRATION

It is a familiar fact that all low pressure manometers so far devised, with the possible exception of the Knudsen radiometer gauge, require calibration. Thus to make absolute measurements with the manometer detector, the intensity of the parent beam must be known for purposes of calibration. For non-condensable gases, this requires (1) an accurate flowmeter to determine the quantity of gas effusing from the source; (2) a knowledge of the mean free path λ_{p_s} (see p. 21 above, and Chapter 2, p. 76); and in any event (3) the value of κ. The accurate determination of all these factors is not easy, and it is therefore fortunate that, as was remarked on p. 26, relative measurements suffice in most cases.

The Hot Wire Gauge.[a] The hot wire gauge utilises the change of resistance suffered by a hot wire when cooled by a gas at low pressure, the resistance of the wire being measured in the standard way by making it one arm of a Wheatstone bridge.

APPROXIMATE THEORY

The energy supplied to the wire by the heating current is dissipated (1) by conduction at the ends; (2) by radiation;

[a] See Table II.

(3) by conduction by the gas. The sensitivity of the gauge is increased in proportion as (3) predominates. (1) is diminished by making the wire long and the cross section small. If for a given cross section the surface is made as large as possible: that is, if the wire is given the form of a thin ribbon, (3) is favoured. It is true that radiation is thereby increased, but since it varies as the fourth power of the temperature, it can be largely suppressed in favour of (3) by lowering the temperature of the wire. With a sufficiently long ribbon-like wire, the end effect (1) can become quite small, and the wire may be treated to a first approximation as infinitely long.

The energy dissipated per unit length of wire per second by radiation is

$$\epsilon\sigma\left(T^4 - T'^4\right)s,$$

where T is the temperature of the wire, T' that of the manometer walls, s the surface per unit length, ϵ the emissive power of the wire, and σ the radiation constant.

The energy dissipated by conduction by the gas is

$$\tfrac{1}{4}n\bar{c}a.\frac{C_v}{N_0}\left(T - T'\right)s,$$

where a is the thermal accommodation coefficient, C_v the specific heat of the gas per mole at constant volume, and the other symbols have their previous significance.

If e is the energy per unit length supplied to the wire, regarded as constant over the small pressure range of an actual experiment, then for temperature equilibrium

$$e = \epsilon\sigma\left(T^4 - T'^4\right)s + ab\left(T - T'\right)s.p,$$

where, by (1·5), $b = \dfrac{C_v}{\sqrt{2\pi R \overline{M} T'}}$

and

$$\frac{dT}{dp} = -\frac{ab\left(T - T'\right)}{4\epsilon\sigma T^3 + abp} = -\frac{ab\left(T - T'\right)}{4\epsilon\sigma T^3} \text{ when } p \text{ is small}$$

$$= -\frac{ab}{4\epsilon\sigma}\cdot\frac{1}{T^2} \text{ when } T \gg T'.$$

Hence
$$\frac{d\rho}{dp} = \frac{d\rho}{dT} \cdot \frac{dT}{dp} = -\alpha_0 \cdot \frac{ab}{4\epsilon\sigma} \cdot \frac{1}{T^2},$$

where ρ is the resistance of the wire per unit length, and α_0 is the temperature coefficient of the resistance.

Or, writing in the value of b in full,

$$\frac{d\rho}{dp} = -\alpha_0 \cdot \frac{\alpha}{\epsilon} \cdot \frac{C_v}{\sqrt{2\pi RMT'}} \cdot \frac{1}{4\sigma T^2} \quad \ldots\ldots(1.12),$$

an approximate equation from which the conditions for high sensitivity may be read.

CONDITIONS FOR HIGH SENSITIVITY

1. C_v and M, constants of the gas, are not at our disposal, but must nevertheless be taken into account when comparing the sensitivity attainable with different gases.

2. The material of the wire must be such that (a) α_0 is large, (b) ϵ is small, i.e. the surface is highly polished. A very highly polished surface may, on the other hand, tend to diminish the accommodation coefficient α.[a]

3. The heating current ($i \propto \sqrt{T}$) must be small. A lower limit is, however, set to i because

(a) The heating current is normally used as measuring current in the bridge; and the bridge sensitivity $\propto i$ (optimum resistance ratios assumed). Thus there must exist an *electrical* optimum for i, when the two effects of increasing gauge sensitivity and decreasing bridge sensitivity with diminishing i balance.

(b) The *thermal* lag of the gauge increases with decreasing T.

Thus the conditions for optimum heating current for a particular set-up are complicated, and are best found by actual trial.

4. The temperature T' of the manometer walls must be low. Two factors set a lower limit to T'.

(a) Condensability of the gas.

[a] See Knudsen, *Ann. Physik*, **34**, 593, 1911.

(b) Thermal transpiration.[a] The receiving slit and lead-out tube are most conveniently held at room temperature T_0. The density of the gas in the manometer is then $\sqrt{T_0/T'}$ that at T_0. The volume V of the manometer vessel is thus effectively increased by a factor $\sqrt{T_0/T'}$; since by (1·11) $t \propto V$, the time lag of the gauge can be inconveniently large for small T'.

The construction of a manometer designed to meet the requirements of high sensitivity is shown in the scale drawing of Fig. 7. The wire is nickel filament of 15μ diameter, rolled to a ribbon of 50μ wide and 4μ thick. Nickel is chosen because (1) it is mechanically strong, (2) it is readily outgassed, (3) it has at the same time a high α_0 value. The resistance of the filament is about 50 ohms. The volume of the gauge (ca. 20 c.c.) is small, to reduce the time lag.

FLUCTUATIONS

The attainment of high sensitivity implies liability to stray fluctuations. Four chief causes of fluctuation fall to be eliminated.

1. Temperature Fluctuations. A second manometer, as far as possible identical with the measuring manometer, and communicating with the observation chamber, is introduced into the neighbouring arm of the bridge. The compensation manometer hangs beside the measuring manometer in the low temperature bath; both are thus subject to the same temperature fluctuations. It is important for temperature compensation (a) to have the glass feet carrying the wire cut from the same length of selected glass rod, (b) to have the lead-in wires (platinum-iridium) go through the pinch without a spot-weld joint.

Fig. 7

[a] See Knudsen, *Ann. Physik*, **31**, 205, 1910.

The presence of the compensation manometer in the bridge of itself increases the sensitivity of the arrangement.[a]

2. Pressure Fluctuations. See p. 35 above.

3. Mechanical Vibration. The purpose of the spiral spring in Fig. 7 is to eliminate disturbances from this source. In later models, it has been found better to discard the spring, and to rely simply on the elasticity of the platinum-iridium leads to take up shock.

4. Electro-magnetic Disturbances. The effect of electro-magnetic disturbances on a normal Wheatstone bridge is absolutely negligible; the manometer wires, however, are deliberately designed to be temperature sensitive and respond instantly to variations of E.M.F. in the bridge circuit. The leads must therefore be made as short and non-inductive as is practically possible; if necessary, they may be run in lead tubing. It may be necessary to surround the manometers with an earthed shield, although usually a commercial Dewar vessel in its metal case, used to cool the manometer walls, suffices.

SENSITIVITY AND REPRODUCIBILITY

For small pressures, the galvanometer deflection is directly proportional to the pressure (cf. p. 39 above). In molecular ray work therefore the galvanometer deflection is read directly, at timed intervals, with the beam alternately cut off by a shutter and admitted to the gauge (see p. 35 above). With equal resistances (ca. 50 ohms) in all four arms of the bridge, and with a galvanometer having a resistance of 60 ohms, and a sensitivity of some 3×10^{-9} amp. per scale division at 1 metre, the deflection at 2 metres for a manometer with the characteristics given on p. 37, is for hydrogen about 1–2 mm. for 10^{-8} mm. alteration in the pressure, the average deflections being constant to within less than 1 mm. With full sensitivity, the parent beam gives therefore a deflection of several metres.

[a] Hale, *Trans. Amer. Electro. Chem. Soc.* 20, 243, 1911.

Fig. 8 shows a point to point plot of a hydrogen parent beam made by moving the detector slit in steps of 0·01 or 0·02 mm. through the beam (see below, p. 57). The circles and crosses represent determinations made on two consecutive days. The small "tails" on either side of the observed

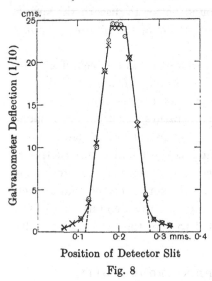

penumbra are due to scattering in the observation chamber, and are difficult to eliminate.

Position of Detector Slit

Fig. 8

It should be mentioned that before Knauer and Stern developed the gauge in its present form, the literature gave 10^{-5} mm. as the minimum pressure change measurable with it; it will be seen that the increase in sensitivity achieved by attention to numerous small details is surprisingly great.[a]

The Ionisation Gauge. The ionisation gauge has been used by Johnson[b] and by Knauer[a] to detect beams of mercury. The pressure p in millimetres of mercury is given by

$$p = k \cdot i/e,$$

where i is the ionisation current given by the electron current e, and k is a constant. The following data, kindly furnished me by Dr Knauer, will give an idea of the order of sensitivity attained. k was found to be 0·1; e was held at $1·5 \times 10^{-2}$ amp. With a pressure of 10^{-6} mm. in the gauge, i/e was 10^{-5}; that is, $i = 1·5 \times 10^{-7}$ amp. With a galvanometer sensitivity of 3×10^{-9} amp. per mm. at 1 metre, this corresponds

a See Knauer and Stern, *Z. Physik*, **53**, 766, 1929 (U. z. M. 10).

b Johnson, *Phys. Rev.* **31**, 103, 1928.

to a deflection of 50 cm., assuming a κ-factor of unity. The ionisation gauge is thus a very sensitive detector for molecules with a low ionisation potential; it is of course not suitable for use with the lighter gases. It has the disadvantage that it cannot readily be compensated for variations in the remanent pressure in the observation chamber, and is therefore very subject to zero drift. Nevertheless, it has probably not received sufficient application to metallic beams.

The Surface Ionisation Detector

According to Langmuir and Kingdon[a] every caesium atom striking a hot tungsten wire ($T > 1200°$ K.) gives up an electron to the wire and evaporates as a positive ion. Potassium and rubidium behave similarly. If the wire is surrounded by a cylinder at negative potential the saturation positive ion current between wire and cylinder gives directly the number of atoms striking the wire per second.

Provide now the cylinder with a suitable window, and place the system in a molecular beam of alkali atoms; then, knowing the dimensions of the wire, one measures *absolutely* and with a time lag depending on the galvanometer characteristics alone the intensity of the beam.

The method is, next to the condensation target, the most sensitive so far devised. Thus taking 2×10^{14} atoms/cm.^{2} sec. as a conservative figure for the intensity at 10 cm. of a beam 1 cm. high, there impinge on a 1 cm. length of wire 0·05 mm. in diameter $5 \times 10^{-3}.2 \times 10^{14} = 10^{12}$ atoms/sec. The corresponding positive ion current is thus

$$10^{12}.1·6 \times 10^{-19} = 1·6 \times 10^{-7} \text{ amp.}$$

The method was worked out by Taylor[b] in Stern's laboratory, initially with potassium as test substance. The wire was of thorium-free tungsten, 0·05 mm. diam., kept taut when hot

[a] For references, see Taylor, *Z. Physik*, 57, 242, 1929 (U. z. M. 14).
[b] Taylor, *loc. cit.*

($T = 1600°$ K.) by a delicate spiral spring; the negative elec-
trode (saturation current at about 10 volts P.D. between wire
and cylinder) was a nickel sheet cylinder, 2 cm. diam., having
two diametrically opposite holes, one to admit the beam, the
other to observe the position of the wire. The whole system
(surrounded except for the necessary loopholes by a liquid air
cooled shield to keep grease vapours, etc., from the wire) was
carried eccentrically on a ground joint, by turning which the

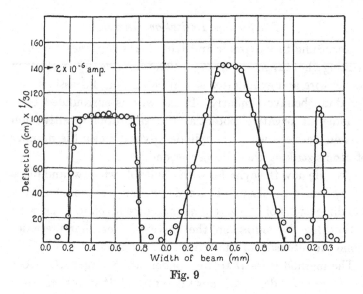

Fig. 9

wire could be traversed through the beam. In Fig. 9 are seen
point to point plots made by Taylor with this method.[a]

The phenomena of surface ionisation on which the method
rests are briefly these: If the electron affinity of an impinging
atom (ionisation potential) is less than that of the surface
(work function), the chances are large that it loses a valence
electron to the surface. Its subsequent history depends on the
temperature of the surface: either it remains adsorbed, at low
temperatures, or at high temperatures evaporates as a neutral

[a] See Taylor, *Phys. Rev.* **35**, 379, 1930.

atom or positive ion. The ratio of the number of ions $\underset{\rightarrow}{s^+}$ to the number of neutral atoms $\underset{\rightarrow}{s^\times}$ leaving unit area of surface per second is

$$\frac{\underset{\rightarrow}{s^+}}{\underset{\rightarrow}{s^\times}} = e^{-F(I-\phi)/RT} \qquad \ldots\ldots(1\cdot13),$$

where I is the ionisation potential of the atom, ϕ the work function of the surface, and F the Faraday number.[a] Thus for $I - \phi \sim - 0\cdot5$ volt, $\underset{\rightarrow}{s^\times}$ is negligible; and, writing dashed symbols to refer to the positive ion saturation current,

$$\underset{\leftarrow}{s^\times} = \underset{\rightarrow}{s^{+\prime}} = \frac{i^{+\prime}}{F} \qquad \ldots\ldots(1\cdot14),$$

which is the basic relation for the detector.

Thus a pure tungsten wire ($\phi_W = 4\cdot48$ v.) at say $1600°$ K. can be used to detect beams of K ($I = 4\cdot1$ v.$_{obs}$), Rb ($4\cdot10$ v.$_{obs}$), Cs ($3\cdot9$ v.$_{obs}$). The importance of having the wire absolutely thorium-free ($\phi_{W, Th} = 2\cdot6$ v.) will be evident.

The range of the method may be extended to include Na ($5\cdot13$ v.$_{obs}$) and Li ($5\cdot37$ v.$_{calc}$) by using an oxygenated wire, which may be prepared very easily by heating the wire *in situ* to $1500°$ K. either in the residual vacuum ($0\cdot1$ mm. to $0\cdot01$ mm.) obtained by simple pumping, or with a pressure of perhaps $0\cdot1$ mm. of pure oxygen in the apparatus. The wire must not thereafter be heated above $1600°$ K., at which temperature the oxygen layer evaporates. Thus Taylor found that a lithium beam, undetectable with a pure tungsten wire, gave the calculated $i^{+\prime}$ with an oxygenated filament at $1500°$ K. If the oxygenated wire were flashed at over $1600°$ K., $i^{+\prime}$ fell immediately to zero.

The value of $\phi_{W, O}$ is unfortunately not known with any

[a] Following Schottky, Wien-Harms' *Handbuch der Experimentalphysik*, vol. XIII, we denote by $\underset{\leftarrow}{s}$, $\underset{\rightarrow}{s}$ the number of particles arriving at or leaving the surface respectively.

certainty,[a] but following Taylor it must lie above 5·4 volts. The possible range of the method is on this count alone—the uncertainty in the $\phi_{W,0}$ value—uncertain. It is, however, not evident *a priori* that (1·14) applies outside the alkali group, within which the atomic core is an inert gas configuration. When this is not the case, it might prove that the chemical forces are so great that the evaporation of the positive ions at temperatures less than 1600° K. is precluded. These points would probably repay further research.

Thermopile

Even the most sensitive thermopile is of little value as a detector of the energy transported by the beam. Since the sensitivity of the pile will clearly depend on the temperature difference between source and detector, it is not possible to give data which will be even approximately applicable to every molecular species. We may take it, however, that an area of 10^{-3} cm.² receives energy from an average beam at 10 cm. from the source at the rate of some 10^{-9} cal./sec. Thus, using for example a Moll vacuum thermo-element (sensitivity, 10^{-8} cal./sec. per microvolt) in conjunction with a Zernicke galvanometer (volt sensitivity, 34 mm. per microvolt at 1 metre), the deflection at one metre is only of the order of a few millimetres.

On the other hand, for atomic hydrogen or oxygen, which have heats of combination of the order of 10^5 cal./mole, the same arrangement yields a deflection of the order of a metre. A thermopile is therefore an excellent detector for even minute intensities of these special rays.

[a] Kingdon (*Phys. Rev.* 24, 510, 1924) determined $\phi_{W,0}$ at 9·2 volts. Later measurements of the contact potential of W, OW by Langmuir and Kingdon (*ibid.* 34, 133, 1929) are in sharp disagreement with this value. It may be that Kingdon's experimental procedure, whereby oxygen was *continuously supplied* to the wire to keep it covered at the temperatures of the experiments, is open to objection.

Bolometer

The sensitivity attainable with a bolometer is about the same as that attainable with a thermopile, and the bolometer used directly as a means of measuring the energy of the beam is therefore of little value. An ingenious application of the bolometer to measure the *heat of condensation* of the molecules from beams of suitable substances has, however, been made recently by Wohlwill, working in Stern's laboratory. I am indebted to Herr Wohlwill for the following particulars of his method.

If Q is the quantity of heat per cm.^{2} per sec. received by the detector, then

$$Q = I.U_c,$$

where U_c is the heat of condensation per mole, and I is the beam intensity. Assuming an intensity of 3×10^{-10} moles/cm.^{2} sec. at 10 cm., we have, since U_c is of the order 10^4 cal./mole, $Q = 3 \times 10^{-10}.10^4 = 3 \times 10^{-6}$ cal./cm.^{2} sec. If the bolometer wire, to reduce heat capacity and conduction, be given the form of a strip, say $50\mu \times 3\mu$, then the receiving area for a beam 1 cm. high is 5×10^{-3} cm.^{2}; and the quantity of heat received is $1 \cdot 5 \times 10^{-8}$ cal./sec. This is about ten times that furnished by the direct energy of impact.

Nickel is chosen as the material for the receiving strip, as having together with high tensile strength a large temperature coefficient of resistance $\alpha_0 = 5 \times 10^{-3}$; with a strip 4 cm. long ($\Omega \sim 10$ ohms at $-180°$ C.), the temperature difference between the middle of the strip, where the beam impinges, and the ends is, very approximately, 7×10^{-2} degrees, giving $\Delta\Omega = 1 \cdot 8 \times 10^{-3}$.

The detecting strip forms one arm of a Wheatstone bridge, the measuring current of which is limited by the fact that the temperature of the strip must lie well below the condensation temperature of the beam substance. Assuming equal resistances (ca. 10 ohms) in all four arms of the bridge, it is in the region of 10^{-2} amp. for substances requiring the strip held at liquid air temperature. With a Zernicke galvanometer, resistance 20 ohms, current sensitivity 4×10^{-10} amp. per mm. at 1 metre, the deflection at 1 metre for the total beam is about 40 cm. The natural lag of the detector is about 30 seconds.

The detector is subject to various disturbances.

1. The oven emits radiation over the whole of the front face. This radiation is received in the plane of the detector over a wide lateral range, many times the width of the beam, and forms a background against which the beam proper must stand out. The heat Q_R received per cm.^{2} per sec. from this source at a distance r is approximately

$$Q_R = \frac{\frac{1}{2}\sigma T^4}{\pi r^2} . A . (1 - \epsilon),$$

where σ is the radiation constant ($1 \cdot 38 \times 10^{-12}$ cal. cm.$^{-2}$ sec.$^{-1}$ degree^{-4}), T is the temperature, A the area of the oven face, assumed half black, and ϵ is the emissive power of the detector: for nickel $\epsilon = 0 \cdot 90$. With $T = 400°$ K., $A = 1$ cm.^{2}, $Q_R = 5 \cdot 6 \times 10^{-6}$ cal./cm.^{2} sec., which is to be compared with the heat $Q = 3 \times 10^{-6}$ cal./cm.^{2} sec. received from the beam. This disturbing factor is readily eliminated by incorporating the comparison resistance of the bridge into the apparatus as a second strip, identical so far as possible with the detecting strip, and held some 2 mm. to one side of it, so as to clear the beam or any deflection pattern it may be desired to plot. In this way, the heat radiation is automatically compensated.

2. Much more serious is the black body radiation *from the oven slit itself*, which cannot be compensated since its lateral range in the plane of the detector is of the same order as that of the molecular beam. In the above example, with a slit 1 cm. long and $0 \cdot 01$ mm. wide, it amounts to only some $5 \cdot 6 \times 10^{-9}$ cal./cm.^{2} sec. at the detector. It can, however, become important at higher oven temperatures. Wohlwill eliminates it by making a preliminary point to point plot with the detector of the cross section of the heat beam from the

slit *of an empty oven* held at the temperature of the subsequent experiment. The deflections so obtained are then subtracted from those observed with the molecular beam, at corresponding settings of the detector. The measurements become awkward, however, when $Q_R > \frac{1}{10}Q$, almost impracticable if $Q_R > Q$.

3. Thermal effects at the connections near the cooled detector are rather troublesome at the present tentative stage of the design; but they could certainly be rendered negligible by a very careful lay-out of the detector.

Radiometer

Radiometers with a reasonably short period of several seconds can be constructed with a sensitivity of some 1/10 radian deflection per microdyne. A parent beam 1 cm. high exerts at 10 cm. from the source a pressure of some 10^{-3} dynes/cm.^{2} Thus the force on a receiving area of 10^{-3}cm.^{2} is some 10^{-6} dynes, corresponding to a deflection of $0\cdot1$ radian, or 10 cm. at 1 metre. The radiometer is thus a practicable direct detector of the beam momentum, but is clearly relatively insensitive.

Certain other means of detection have presented themselves from time to time as possibilities to workers in the field of molecular rays, but the discussion has been confined here to detectors which have actually been tried out in practice. For the purpose of easy reference the detectors discussed here, with the substances to which they have been applied, and the relevant sources of information *in so far as they deal with the methodics*, are collected in Table II. Those marked with an asterisk have not received any special development. It will be seen from the Table that there exists no *quantitative* method for the detection of beams of the heavier metals and the less volatile compounds. This lack is at present the most serious gap in the technique.

Table II

Detector	Substances studied	Literature (methodics)
1. Condensation target	Li, Na, K; Cu, Ag, Au; Zn, Cd, Hg; Tl; Sn, Pb; As, Bi; Mn; I; Fe, Co, Ni H_2O, N_2O_5; CsCl, RbBr; NaI, KI, TlI (benzene ring)–CH_2–(benzene ring), (benzene ring)–O–(benzene ring), (benzene ring)–NH–(benzene ring), (benzene ring)–CO–(benzene ring) $o, m, p - C_6H_4.NH_2.COOCH_3$ $C(CH_2OH)_4$ and derivatives Pinene	General: Knauer and Stern, Z. Physik, **39**, 764, 1926. Developers: Method I. Estermann and Stern, Z. physikal. Chem. **106**, 399, 1923; Gerlach, Ann. Physik, **76**, 163, 1925. Method II. Knauer and Stern, loc. cit. Weighing: Ellett, Olson and Zahl, Phys. Rev. **34**, 493, 1929; Wulff, Z. physikal. Chem. B, **6**, 43, 1929. Wiener Method: Bielz, Z. Physik, **32**, 81, 1925. Johnson, J. Franklin Inst. **207**, 629, 1929; ibid. **210**, 141, 1930 (MoO_3 target).
2. Chemical target	H, O, active N, HCl	Knauer and Stern, Z. Physik, **53**, 766, 1929. Johnson, Phys. Rev. **31**, 103, 1928.
3. Manometers: (a) Hot wire gauge (b) Ionisation gauge	H_2, O_2, CO_2, He, Ne, Ar Hg	Taylor, Z. Physik, **57**, 242, 1929; Phys. Rev. **35**, 375, 1930.
4. Surface Ionisation detector	Li, Na, K, Cs	Knauer (unpublished).
5. Thermopile	H	
6. *Bolometer Condensation bolometer	H_2, He $C(CH_2OH)_4$, etc.	Solt, Phys. Rev. **29**, 904, 1927 (Abs.). Wohlwill (unpublished).
7. *Radiometer	H_2, N_2, CCl_4	Costa, Smyth and Compton, Phys. Rev. **30**, 349, 1927.

CONSTRUCTIONAL DETAILS

The total path of the beam from source to detector will not in general exceed some 20 cm. If the path is made much longer, the intensity of the beam at the detector becomes inconveniently small; and moreover, much greater attention must be paid to the vacuum conditions in the apparatus, which in the ordinary way merely demand a maximum pressure of some 10^{-5} mm. Undoubtedly the correct line of attack in attempting to gain sensitivity in experiments on the deflection, scattering, reflection, and so forth of molecular

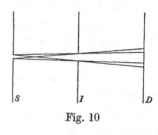

beams is to work with the narrowest possible slits rather than with a long path.

The dimensions of the cross section of the beam at the position of the detector are determined by the size of the slits and their distance apart, in accordance with the laws

Fig. 10

of geometrical optics. Fig. 10 shows a cross section parallel to the slit width for the case, image slit twice the width of source slit, and $SI = ID$. In practice it is desirable to suppress the penumbra as far as possible, so that SI cannot in the attempt to gain intensity be made too small. As a rough rule, $SI = 1/3\ SD$ is satisfactory.

When the length of path $SD = r$ enters in the final result (as for example in the magnetic deflection experiments, see Chapter 5), the error arising from the finite length of the slit must be less than the error of observation. If $h = 2l$ is the length of the slit, the path lengths of molecules arriving at the centre of the detector lie between r and

$$\sqrt{r^2 + l^2} = r \left(1 + \frac{1}{2} \frac{l^2}{r^2} \right)$$

for small l/r. The error is thus $1/2 \cdot l^2/r^2$. It frequently happens that the conditions of experiment of themselves limit the length of the slit. Thus in magnetic measurements $h \sim 2$ mm.; with $r = 10$ cm., $1/2 . l^2/r^2 = 5 \times 10^{-5}$, well within the experimental error.

Adjustment. The slits are normally formed of adjustable jaws running in dovetailed slides, except for high temperature ovens, when they are usually screwed on.

The simplest method of adjusting the slits parallel is with telescope and crosswire. The distance SI, about 4–6 cm., pre-

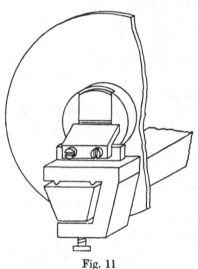

cludes high magnification, and the method is not of the highest accuracy for long slits. Where the slit length is limited to say 2 mm. (see p. 25 above), the method is sufficiently exact; for although the effective length of the slit is limited by suitable diaphragms to 2 mm., the adjustment can be made on the slit jaws, which are perhaps 5 mm. in length.

Fig. 11

When the source temperature is not too different from room temperature, an extremely reliable procedure is the following. The slits are mounted on a metal bar, like an optical bench, with a profile like that seen in Fig. 11. They are adjusted parallel to the upper surface of the bench (which should be ground flat to 10^{-3} mm.) by means of a rider, which is furnished with a knife edge previously aligned parallel to the bed (see Fig. 11, which is self-explanatory). Under the microscope, the slits can be adjusted parallel to the knife edge to within a few microns.

Where a fore slit (see p. 22 above) is used, the short distance (ca. 4 mm.) between source slit and fore slit presents a difficulty. This can be successfully overcome by mounting the source slit on a detachable carrier, which is held down on the bed on three points by means of a single screw. The adjustment of the source slit is carried out with the help of the rider on a subsidiary bench, the carrier is then transferred to its place on the bench proper, and screwed down not too forcibly into position.

The fore slit should be made much wider than the source slit, say ten times as wide, when slits of the order of 0·01 mm. are used. Its function being merely to step down the pressure in the collimator chamber, there is no objection to this. The precaution is necessary because source slit and fore slit being separated by a few millimetres only, a misadjustment by only a few microns of a fore slit of 0·01 mm. is sufficient to wreck the beam.

It is better to make the image slit rather wider than the source slit. Otherwise there ensues a loss in intensity, due probably to the increased difficulty of adjustment; or in case a feeble oven cloud is present, to a decrease in the area of the cloud available for beam formation.

A weakness common to all methods of adjustment is the following: the slits are adjusted parallel at room temperature; under the conditions of experiment their temperatures may differ by hundreds of degrees. For narrow slits, this effect may become important. Knauer and Stern[a] showed that this difficulty may be overcome by controlling and correcting the adjustment under the conditions of experiment by means of the rays themselves. To this end, the middle zone of the source slit is covered with a strip of platinum foil, so that only the two ends emit molecules. Each end then throws its own image of the image slit on the plane of the detector. If the slits are parallel the two images coincide; if not they fall side by side. From their distance apart (or near perfect adjustment, from the width of the image) the departure from parallelism can be

[a] Knauer and Stern, Z. Physik, 39, 771, 1926 (U. z. M. 2).

calculated, and the image slit rotated by the required amount. Two or three trials are usually sufficient to secure perfect alignment.

Typical Apparatus. Two typical pieces of apparatus, the first using the condensation method, the second the pumping method of eliminating alien molecules (see p. 20 above) will

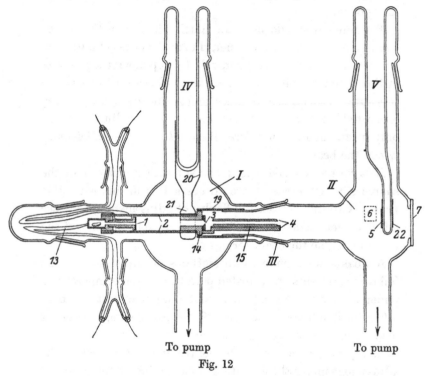

To pump   To pump
Fig. 12

now be described in detail. In this way the salient points in the constructional technique are automatically covered.

Fig. 12 shows an apparatus used by Estermann[a] for the determination of the molecular electric moments of easily condensable substances. The arrangement for producing an inhomogeneous electric field is seen at (15), and does not concern us further here.

[a] *Z. physikal. Chem.* B, 1, 161, 1928.

The apparatus, of glass, is in two parts, the collimator
chamber I and the observation chamber II, connected by the
ground glass joint III. The male cone carries on a cylindrical
extension a well-fitting metal mantle (19) into which the
image slit carrier (14) can be screwed. I is in this way entirely
separated from II, except for the image slit (3); stray molecules from I are thus prevented from reaching the sensitive
target (5). I and II are evacuated by separate pumps.

The molecule ray gun is shown hachured in the figure. It
is supported at the rear of (14), which is oval in section, by
the elastic brass fork (21), the prolongation of a cylinder
fitting snugly over the Dewar vessel IV. The molecules issue
from the oven slit (1); a beam is selected by the image slit
(3), which is rigidly attached to the oven by the constantan
strips (2), and is received on the polished nickel target (5),
cooled to about − 160° C. by the Dewar vessel V. The resulting deposit is observed microscopically through the plane
glass window (7) by means of the totally reflecting prism (6).
Alien molecules are removed by condensation on the large
cooled surface of IV.

Details of the oven construction are seen in the small sketch
(Fig. 12 a). The oven consists of
an axially bored phosphor-bronze
block, into the front of which is
screwed the oven slit carrier (9);
the substance container (11) is a
push-in fit into the back. The oven
is heated by the furnace (10), which
consists of a hard glass tube platin-
ised to give a resistance of a few ohms.[a] The phosphor-bronze

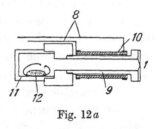

Fig. 12 a

[a] This form of furnace is very handy for oven temperatures up to say
200° C. The glass is platinised by painting on "Westhaversche Platinlösung"
(supplied by Heraeus Hanau) thinly and uniformly with a camel's hair
brush, then burning off the lavender oil by heating (keep turning!) to a dull
red heat. By varying the number (*not* the thickness) of the coats, and by
varying the degree of "burning in", a film of the required resistance can be

block is carried by the constantan strips (2), which are cooled by IV. Thus the oven slit is the hottest part of the oven, an important point in the design, preventing as it does deposition of substance in the slit. The temperature is measured by the copper-constantan thermo-element (Figs. 12, 13). The current leads (8) can also be seen in the figure.

It is important to have good thermal contact between the metal parts (20) and (22) and their respective Dewar vessels. This is very successfully achieved by platinising the glass (which must be hard glass, to avoid deformation) and electro-depositing copper on the platinum surface to a thickness of a few tenths of a millimetre. The copper coat can be turned accurately cylindrical in the lathe, and a good sliding fit for (20) and (22) secured.

Fig. 13 shows the apparatus used by Knauer and Stern[a] to determine λ for hydrogen (see Chapter 2, p. 76). The source slit O, to which the gas is led via a capillary from a reservoir, the image slit Ab, and the detector slit Af, communicating with a hot wire gauge, are mounted rigidly on a steel bench, seen in cross section in the small sketch. They are adjusted parallel to the upper surface of the bench by means of a rider (see Fig. 11). S is a magnetically operated shutter. The image slit is carried on a brass disc, which is screwed to a ring let into the cylindrical brass housing; this serves the double purpose of supporting the slit system, and together with the glass end plates, of dividing the housing into two compartments, collimator chamber and observation chamber, which communicate separately each with a large Leybold steel pump. It will be observed that the apparatus, intended as it is for mean free path determinations (see p. 24), does not include a fore slit.

Detector Slit. The chief point of interest in the apparatus is

got. The ends to a width of a few millimetres are more thickly coated and less strongly burned in, to give good contact with the current leads.

[a] *Z. Physik*, 53, 766, 1929 (U. z. M. 10).

the detector slit Af. The detector slit carrier is a push-in fit into a dovetailed slide which is moved against a leaf spring (not shown) across the beam from outside the apparatus by means of the screwdriver shown. The position of Af is read off on the graduated drum; with a pitch of $\frac{1}{2}$ mm. to the screw and a well-sized drum, easily to 5μ.

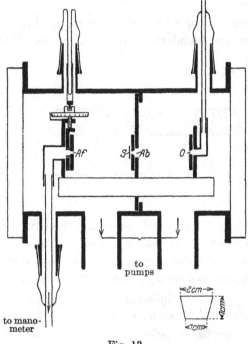

Fig. 13

Typical dimensions for the detector slit are given in the profile sketch, Fig. 14 a. The slit is given the form shown for two reasons. (1) It defines the permissible error for the angle made by the axis of the beam with the axis of the canal to $\theta = 1/30$ radian. (2) It allows of a traverse of 2–3 mm. at right angles to the beam; this is particularly necessary, as it is much less convenient to make the detector slit move over the arc of a circle, so as always to present itself at right angles to the beam.

The stationary pressure (see p. 34 above) establishes itself initially in the space s immediately behind the slit (Fig. 14 b). It is found that the time lag is noticeably shortened if a channel is cut in the base of the carrier parallel to the slit length, debouching at the centre into the lead-out tube to the manometer.

If after the slit edges have been ground parallel, the corners of the jaws are found to be rounded, as indicated by the

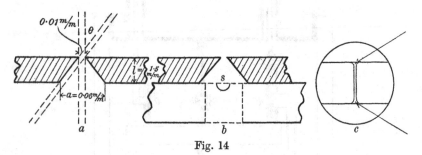

Fig. 14

arrows in Fig. 14 c, two little canals are thereby formed from s into the observation chamber which are equivalent to a leak in the manometer system. The holes can be stopped up with a little stiff tap grease which has been previously melted *in vacuo* to remove occluded air.

The κ value of a slit such as that shown in the figure can be calculated from the equation[a]

$$\kappa = \frac{3}{8}\frac{l}{a} + 1 \qquad \ldots\ldots(1\cdot15).$$

[a] The values of $\kappa = \dfrac{R_c + R_s}{R_s}$ calculated from equation (1·15) agree to within 10 per cent. with those determined experimentally, in spite of the fact that (1·15) is derived from the incorrect Knudsen formula for the resistance of a canal of arbitrary shape $R_c = 3/8 . \sqrt{\pi/2} \int_O^L O/a^2 . dl$, where O is the periphery, a the area of a cross section of the canal (Knudsen, *Ann. Physik*, **28**, 76, 1909; see also Smoluchowski, *ibid.* **33**, 1559, 1910).

Fore Slit. Where a fore slit is used, the source slit may be given a V-shaped form to diminish the Knudsen resistance in the narrow space between it and the fore slit. If the latter is also made V-shaped so much the better; the effect of the directly reflected alien molecules (see p. 22 above) is thereby also diminished. Although several designs have been tried out with fair success a completely satisfactory form for these slits has not so far been found.

Chapter 2

GAS KINETICS

The pioneer experiments of Dunoyer, described in the Introduction, not only demonstrated the accuracy of the fundamental postulates of the kinetic theory, but gave immediate
promise of a wide application of the method of molecular rays
to the experimental study of gases. Dunoyer wrote: "La
possibilité d'observer un rayonnement matériel dont l'énergie
est d'origine entièrement thermique ouvre la voie à un assez
grand nombre de recherches, dont l'ensemble pourrait porter
le nom de *cinétique expérimentale*".[a]

He outlined a programme of research which should bring
molecular quantities within the ambit of direct experiment.[b]
He suggested, first the determination of the mean velocity $\bar{c}$
by measuring radiometrically the momentum $M\bar{c}$ of the beam,
M being the mass of substance deposited from the beam per
second; next the measurement, by bolometer or thermopile,
of the energy $1/2 . MC^2$ transported by the beam. $\bar{c}$ and C
known, evaluation of the ratio $\bar{c} : C = \sqrt{8/3\pi} = 0.921$ would
give a direct experimental verification of the Maxwell law of
the distribution of velocities.[c]

It is clear that the ratio $\bar{c} : C$ would have to be determined
correct to a few per cent. before a reliable comparison with
the theoretical value could be made; and we have seen in
Chapter 1 that the direct determination of the energy and
momentum of the beam to this degree of accuracy is practically excluded. It is therefore not surprising that the direction
actually followed by the technique has been quite another
than that originally suggested. The goal has remained essen-

[a] Dunoyer, *Compt. rend.* 152, 595, 1911.
[b] Dunoyer, *Le Radium*, 8, 146, 1911.
[c] See Appendix.

tially the same: namely, a direct experimental verification of the Maxwell distribution law; but it has been achieved with the aid of mechanical devices which reduce the necessary measurements to a simple comparison of intensities.

THE DIRECT MEASUREMENT OF THE THERMAL VELOCITY OF MOLECULES

A first rough direct measurement of molecular velocities was made in 1920 by Stern,[a] using an ingenious device, the principle of which will be made clear by a study of Fig. 15. At the centre of an evacuated cylindrical vessel V, of radius l, is placed a gas filled vessel O, furnished with a slit S_1. A second slit S_2 defines a molecular beam which is received at the wall of V at the point P. Let the whole apparatus be set in clockwise rotation about an axis through S_1; then during the time $t = l/v$ required for a molecule with velocity v to reach the wall of V, a point in that

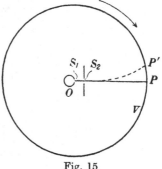

Fig. 15

wall will have described a distance of arc $s = 2\pi n . l . t$, where n is the number of revolutions of the apparatus per second. Thus a molecule which arrived at the point P with the apparatus at rest will, with the apparatus in rotation, strike the outer wall at a point P', whose distance from P, measured in an anti-clockwise direction, is

$$s = 2\pi n . l^2/v \qquad \ldots\ldots(2\cdot1).$$

It has so far been tacitly assumed that the distances S_1P, S_2P are sensibly equal, an arrangement which would preclude the formation of a well-defined beam. If $S_1P = l_1$ and

[a] Stern, Z. Physik, 2, 49, 1920; ibid. 3, 417, 1920.

$S_2P = l_2$ it is easy to show that for small s, l in (2·1) is to be replaced by $\sqrt{l_1 l_2}$. Hence

$$s = 2\pi n \cdot l_1 l_2 / v \qquad \ldots\ldots(2\cdot1').$$

It may be noted that the deflection s is quite easily measur-

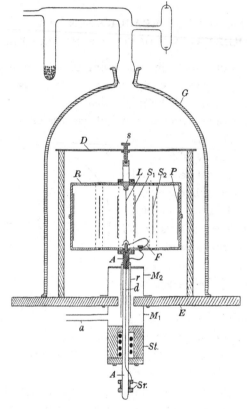

Fig. 16

able; for example, if in (2·1) $n = 50$, $l = 10$ cm.; then for $v = 5 \times 10^4$ cm./sec., $s = 6$ mm.

The actual apparatus as realised by Stern for the study of a silver beam is seen in Fig. 16. The rotating system R is carried inside the evacuated bell-jar G on the axis A, which is led out from the apparatus through a stuffing box $M_1 M_2$, kept constantly evacuated by a pump at a, to the axle of a motor. A second bearing is formed by the pointed adjustable screw s. The source in this case is not a gas-filled vessel, but a silvered platinum wire L carried in the axis of the rotating system. The silver sheath is raised to the melting point by a current passed through the platinum wire, the temperature of which is estimated pyrometrically. The current leads are connected to the two slip rings at Sr; from the lower ring, via the axle and frame R to the upper end of the wire; from the upper insulated ring to the bottom end of the wire via the insulated leaf spring F, which also serves to keep the wire taut when hot. The slit system $S_1 S_2$ and the brass target T are duplicated, one on either side of the wire, to ensure good balance of the rotating parts. The function of the additional slits S_1 is simply to eliminate the effect of a possible movement of the wire during an experiment, for only so long as $L S_1 S_2$ are lined up can a trace appear on the target.

The frame was first ro-
tated clockwise, and a deflected trace obtained; the
experiment was then re-
peated, with equally rapid
anti - clockwise rotation.
Thus two traces, corre-
sponding points of which
are separated by $2s$, were obtained, the accuracy of the
measurement being in this
way considerably increased.
The traces obtained in two

Fig. 17

runs, with 2400 and 2700 revs. per min. respectively, are shown in Fig. 17. It will be noticed that the traces are rather broad and diffuse in outline; this is of course precisely what is to be expected on the assumption of a statistical distribution of velocities among the silver atoms leaving the wire.

The velocity corresponding to the middle of the deflected trace was found from (2·1′) to be 643 m./sec. and 675 m./sec. respectively, or approximately $\sqrt{3 \cdot 5kT/m}$.

THE MAXWELL DISTRIBUTION LAW

Quantitative measurements of the intensity distribution in the deflected traces obtained in Stern's experiment, had they been possible, would have given direct information about the velocity distribution of the silver atoms. For in general, if $dn = f(v)\,dv$ is the instantaneous number of molecules in the source with velocities between v and $v + dv$, the direction of v being within indefinitely small limits that of the axis of the beam; then the number arriving on unit area of the detector per second is

$$I(v)\,dv = vdn = f(v)\,vdv \qquad \ldots\ldots(2\cdot2).$$

The form of $f(v)$ has still to be assigned; if the Maxwell distribution of velocities is assumed, it is of the form $Ce^{-v^2/a^2}.v^2$, where a is the most probable velocity of the molecules in the source. Thus

$$I(v)\,dv = C'e^{-v^2/a^2}.v^3.dv \qquad \ldots\ldots(2\cdot3).$$

The value of C' is found by integration to be $2I_0/a^4$, where I_0 is the total beam intensity for all velocities; whence finally

$$I(v)\,dv = \frac{2I_0}{a^4}e^{-v^2/a^2}.v^3.dv \qquad \ldots\ldots(2\cdot4).$$

Now the deflection s_a in Stern's experiment corresponding to the most probable velocity a is, from (2·1′),

$$s_a = 2\pi n.\frac{l_1 l_2}{a} \qquad \ldots\ldots(2\cdot5).$$

Hence, changing the independent variable in (2·4) from v to $s = s_a.a/v$, we have, assuming that the width of the parent beam is negligible in comparison with that of the deflected trace,

$$I(s)\,ds = 2I_0.e^{-s_a^2/s^2}.\frac{s_a^4}{s^5}.ds \qquad \ldots\ldots(2\cdot6)$$

for the intensity corresponding to deflections between s and $s + ds$.

Lammert's Experiments. The necessity of making a point to point plot of the intensity distribution in a velocity spectrum, which can only be expected to give semi-quantitative results with a target detector, can be avoided by comparing the intensities *of selected velocity ranges* against the total beam intensity.

The method used to select the velocity ranges bears a close analogy to the Fizeau method of determining the velocity of light by means of a rotating toothed wheel.

Imagine to begin with two discs mounted on a common axis, in the first of which is cut at the periphery a single radial slot; on the inner face of the second disc we shall suppose that a radial reference line has been scribed off exactly opposite the slot in the first disc. Suppose that the source slit is placed in a radial direction a short distance from the outer face of the first disc, and opposite the slot. We shall consider only molecules which emerge from the source slit in directions parallel to the axis of rotation of the discs. With the discs at rest, these molecules will strike the second disc at the position of the reference line. Suppose the discs to be set in clockwise rotation; then molecules can pass the first disc only during the short interval of time that the slot is opposite the source slit. Now a molecule which has passed through the slot in the first disc with velocity v arrives at the second disc after a time $t = l/v$, where l is the distance between the discs. During this time the reference line has swung out of the line of fire through an angle $\delta = \omega.t$, where ω is the angular velocity of rotation. Hence a molecule of velocity v will hit the second disc at an angular distance $\delta = \omega.l/v$ counter-clockwise to the reference line. If now a slot is cut in the second disc at this point, of angular width 2γ, then molecules can pass the second disc whose velocities lie between

$$v_1 = \frac{\omega l}{\delta - \gamma} \text{ and } v_2 = \frac{\omega l}{\delta + \gamma}.$$

Thus if it is desired to select a velocity range between v_1 and v_2, the number of revolutions of the discs per second is

$$n = \frac{\gamma}{\pi l} \cdot \frac{v_1 v_2}{v_1 - v_2} \qquad \text{......(2·7)}$$

and the angle of displacement

$$\delta = \gamma \cdot \frac{v_1 + v_2}{v_1 - v_2} \qquad \text{......(2·8).}$$

The intensity I of the velocity range $\Delta v = (v_1 - v_2)$ is from (2·4)

$$I = I_0 \left[e^{-v^2/a^2} \left(1 + v^2/a^2 \right) \right]_{v_1}^{v_2} \qquad \text{......(2·9).}$$

Actually, it is not possible to have only one slot per disc; for clearly the initial intensity is reduced by the first disc in the ratio $n\gamma/\pi$, where n is the number of slots. Hence as many as 50 to 100 slots per disc may be used, the angle of displacement δ being got by giving the second disc a suitable twist relative to the first, in a direction opposite to that of rotation. The molecules arriving at the detector may be confined to those travelling parallel to the axis of rotation by placing an image slit in a radial direction behind the second disc.

The experiment was carried out for a mercury beam by

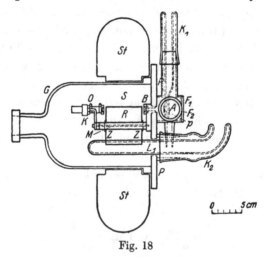

Fig. 18

Lammert, working in Stern's laboratory.[a] His results constitute the first reliable direct experimental verification of the Maxwell distribution law. The apparatus used is seen in Fig. 18. It is divided into collimator chamber S and observation chamber A by the dividing wall P carrying the image slit B. The oven, attached to the constantan strip K, is seen with its platinised glass heating jacket at O, and between oven slit and image slit lie the slotted discs ZZ, forming the ends of an iron cylinder R (diameter 5·7 cm., length 6 cm.) carried on ball bearings, which serves as rotor to the stator St, surrounding the apparatus. The speed of the rotor is determined stroboscopically. The silvered target is at p, cooled by liquid air contained in the Dewar vessel K_1. K_2 is a Dewar vessel giving some 80 cm.[2] of cooled surface in S.

The construction of the rotor deserves a more detailed description. The form of the discs is seen in Fig. 19. The outer half of all the 50 slots is covered by a ring, in which is cut two sectors of total aperture just equal to the aperture of all the uncovered slots together. Then the same fraction of the total beam passes the first disc via sectors or slots; but whereas the whole sector fraction can pass

Fig. 19

the sectors of the second disc, only the desired velocity range can pass its slots. Thus the trace on the target appears in two portions: the upper half, arising from molecules which have passed the sectors, first; the lower half, due to the selected range, considerably later. The ratio t_s/t of their times of appearance equals the ratio I/I_s of the selected intensity I to the sector intensity I_s. In this way the uncertainty inherent in the determination of the time of appearance, as well as possible variations in total beam intensity, are eliminated or reduced to a minimum.

[a] Lammert, Z. Physik, 56, 244, 1929 (U. z. M. 13). Prior to the publication of Lammert's paper, the demoniac behaviour of a similar apparatus was reported by Costa, Smyth and Compton (Phys. Rev. 30, 349, 1927).

GAS KINETICS

The agreement of the observed intensity ratios $I/I_s = t_s/t$ with those calculated from (2·9), which assumes the Maxwell distribution, is good, as may be seen from Table III, taken from Lammert's paper. Five equal velocity ranges of 50 m./sec. were selected by suitably choosing n and δ (equations (2·7) and (2·8)) and the corresponding times of appearance t and t_s observed. The ratios t_s/t are to be compared with the values I/I_s in the last column, which have been corrected for the very slow molecules which, particularly at high speeds of rotation, may get through a slot in the second disc with an angle of displacement $(\theta + \delta)$, θ being the angle between successive slots. Fig. 20 shows the above series of measurements graphically. The areas of the rectangles represent the intensities of the different velocity ranges; full lines correspond to the measured values, broken lines to values calculated from

Table III

No.	v_2 to v_1	n	$\delta°$	I/I_s %	t_s		t		t_s/t %	I/I_s (corr.) %
	140 to 190	14·1	1·89	19·4	min.	sec.		Mean:	19·1	19·5
1					4	30	22 min.	10 sec.	20·25	
2					4	—	22	—	18·2	
3					5	—	27	—	18·5	
4					5	15	27	—	19·45	
	190 to 240	24·2	2·46	23·2				Mean:	23·0	23·5
5					7	30	31 min.	30 sec.	23·8	
6					8	40	40	—	21·65	
7					4	45	20	—	23·75	
8					6	20	27	40	22·9	
	240 to 290	37·0	3·04	19·9				Mean:	19·8	20·4
9					8	20	40 min.	— sec.	20·85	
10					9	—	47	—	19·15	
11					9	40	49	30	19·5	
	290 to 340	52·4	3·61	13·1				Mean:	13·5	14·3
12					5	20	37 min.	50 sec.	14·1	
13					5	—	38	—	13·15	
14					6	—	45	—	13·3	
	340 to 390	70·0	4·18	6·8				Mean:	9·0	9·8
15					4	30	51 min.	45 sec.	8·7	
16					4	45	51	—	9·3	

(2·9). The small systematic difference is probably due to a slight inequality in the apertures of the sectors and slots.

The velocity corresponding to the maximum of the curve $\left(\dfrac{df(v)}{dv} = 0; f(v) = e^{-v^2/a^2} . v^3 \text{ by } (2 \cdot 3)\right)$ is $v = \sqrt{\tfrac{3}{2}}a$, which for mercury at $100°$ C. is 214 m./sec., in full agreement with observation.

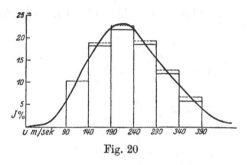

Fig. 20

Eldridge's experiments. An ingenious variant of the slotted disc device, which allows the entire velocity spectrum to be received simultaneously on a *stationary* target, has been used by Eldridge[a] to study the velocity distribution of cadmium atoms at a temperature of $400°$ C. (Fig. 21). It consists essentially of the Lammert arrangement without the image slit. The atoms reaching the target are therefore not confined to those moving parallel to the axis of rotation of the discs, but instead use is made of the *cone* of rays emerging from the slots in the lowest disc, within a semi-vertical angle of about $4°$; within this angle the intensity is constant to within 2 or 3 per cent.

The mode of operation of the device is as follows. If we imagine the intermediate discs removed and only one slot each in the remaining two discs, then the slot in the upper disc sweeps over the cone emerging from the lower disc *in the direction of rotation*, allowing the fastest molecules to pass at

[a] Eldridge, *Phys. Rev.* **30**, 931, 1927.

the beginning of its traverse, and ever slower molecules as the traverse continues. Thus an entire velocity spectrum is sorted out on the stationary cooled target, with the fastest molecules registering most closely to the position on the target opposite the source slit; the displacement s corresponding to a given velocity v being clearly

$$s = 2\pi n . lr/v \qquad \ldots\ldots(2\cdot10),$$

where r is the radius of the disc. When in order to raise the

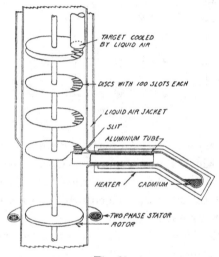

Fig. 21

intensity many slots are used, it must always be ensured that in the upper disc only those slots *following* a certain marked slot allow molecules to pass. This is the function of the intermediate discs; if a marked slot in the upper disc, distant l from the lower, has a displacement δ when passing molecules of velocity v, the intermediate discs are given a twist in the direction of rotation, namely $\delta = \omega l_n/v$, where n is 1, 2 or 3. Thus only those molecules having the desired ray path are allowed to pass. A disadvantage of the arrangement is that the series of slots through which a given ray passes are not

GAS KINETICS71

exactly parallel owing to the rotation of the discs in transit;
this especially affects the molecules with the larger deflections
(that is, the slower molecules). It can be partially compensated by making the slots in the intermediate discs somewhat wider than those of the extreme discs.

Eldridge's results are shown graphically in Fig. 22. The
circles represent the intensities on an arbitrary scale corresponding to different displacements s, as determined by
photometering the deposit. To the left is seen the undeflected
trace, made with the rotor in very slow rotation; to the right

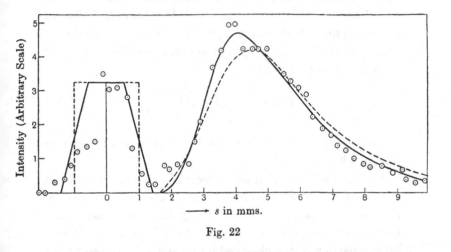

Fig. 22

is the plot of the velocity spectrum. The continuous curve in
the figure is calculated from (2·6), and it will be seen that the
observed points lie on it fairly closely.

The agreement is however fortuitously good, chiefly because (2·6) is applicable to a displacement curve for which
$s \propto l/v$ (cf. (2·10)), only so long as the width of the undeflected
trace is negligible in comparison with that of the deflected
trace: whereas it is clear from Fig. 22 that the undeflected
trace has a width of some 2 mm. as against some 8 mm. for
the deflected trace.

If in general the undeflected trace has a width $2a$, then $I(s)$ in (2·6) represents only that fraction dI of the intensity in the portion ds of the deflected beam which arises from the infinitesimal portion ds_0 of the undeflected beam for which the parent intensity is I_0. That is,

$$dI = 2I_0 ds_0 e^{-\frac{s_a^2}{(s-s_0)^2}} \cdot \frac{s_a^4}{(s-s_0)^5} \qquad \ldots\ldots(2\cdot10').$$

To obtain the total intensity in the deflected beam at the point $s' = s - s_0$, we should, strictly speaking, integrate over I_0 as a function of s_0; that is, we require to know the intensity distribution in the undeflected trace. It is however sufficient for most purposes to assume I_0 constant over the undeflected trace; when, integrating (2·10') between the limits $s_0 = \pm a$ with $I_0 = $ const., we get

$$I = I_0 \left[e^{-y}(y+1) \right]_{\left(\frac{s_a}{s-a}\right)^2}^{\left(\frac{s_a}{s+a}\right)^2} \qquad \ldots\ldots(2\cdot11)$$

for the intensity distribution in a deflected beam arising from a parent beam of width $2a$.

The intensity distribution corrected for the finite width of the undeflected trace is represented on the same scale as the full curve by the broken curve in Fig. 22. The intensity loss suffered by the slower molecules is brought out very clearly on comparison of the observed intensities with the corrected curve.

Apart from the effect of the finite width of the parent beam, which can be accounted for, the results are uncertain in so far as the photometry of a metallic deposit as a means of determining intensities is itself uncertain (see Chapter 1, p. 31), as may indeed be seen from the plot of the undeflected trace in Fig. 22. Thus Eldridge's results cannot be given the same weight as those of Lammert; but the device is undoubtedly capable of refinement, and should prove useful when a rapid survey of the velocity conditions in a beam is desired.

VELOCITY SELECTORS

It is becoming of increasing importance for many applications of the molecular ray method to have the molecules in the beam moving with velocities which lie within a definite narrow velocity range. Such a beam may be called a "monochromatic" molecular beam. This is in no way a misnomer, for a single velocity beam is indeed monochromatic in the de Broglie waves of the molecules. Any device which produces a beam of selected velocity range may be called a "monochromator" or alternatively a "velocity selector".

The slotted disc device, in the form developed by Lammert, is a practicable velocity selector. It has however notable disadvantages. The total beam intensity is weakened by passage through the first disc in the ratio $n\gamma/\pi$, where 2γ is the angular width of a slot, and n is the number of slots: in practice, to about 10 per cent. of the parent intensity. Thus only $0 \cdot 1 I_0$ is available from which to select the desired velocity range. Now a 50-metre range at the most probable velocity amounts to some 25 per cent. of the whole velocity spectrum, so that the selector delivers a maximum of only some $0 \cdot 03 I_0$. It must also be remembered that the inclusion of the selector inevitably lengthens the total path of the beam, which also acts to reduce the available intensity.

The Stern device (see p. 62 above) is in this respect much more efficient, because the total beam intensity is available for selection by a slit placed at the appropriate position on the target. It has the very great disadvantage in practice that the entire apparatus in which it is desired to study the monochromatised beam would have to rotate. It seems probable that the technical difficulties involved will prove too great ever to make of this arrangement a practicable selector.

A velocity selector employing vibrating slits has been described, but not yet tested out, by Tykocinski-Tykociner.[a] It

[a] Tykocinski-Tykociner, *J. Opt. Soc. Amer.* 14, 423, 1927.

does not promise to be a very useful type, because if the distance between the slits is to be reduced to a reasonable figure, the frequency of vibration must be very high. It has the further disadvantage that there are *two* factors, the frequency and the amplitude of vibration, to keep constant, whereas the slotted disc selector requires only one factor to remain steady, namely the speed of revolution.

The method of monochromatisation which will probably find the widest application in the future, at any rate to the lighter gases, makes use of the wave nature of matter. The desired de Broglie wavelength is selected from the diffraction spectrum obtained by reflecting a molecular beam from the surface of a crystal, in precise analogy with optical practice (see Chapter 4). The absence of moving parts is of course an enormous advantage. Monochromatic beams of hydrogen and helium have already been studied successfully by this method in Stern's laboratory.

THE MEAN FREE PATH

The first rough measurements of a mean free path by a molecular ray method—that of silver in nitrogen or air—were made by Born[a] and Bielz[b] in 1920–25. Suppose that with a perfect vacuum the amount of silver deposited by the beam in a given time t on a surface distant l from the source is M_l^o; then the amount deposited when the gas pressure in the apparatus is p is

$$M_l^p = M_l^o \cdot e^{-l/\lambda} \qquad \ldots\ldots(2\cdot12),$$

where λ is the mean free path. Comparison by special methods of the amounts of silver $M_{l_1}^o$, $M_{l_2}^o$; $M_{l_1}^p$, $M_{l_2}^p$ deposited on two small surfaces distant l_1, l_2 from the source, and which are exposed to non-overlapping sectors of a silver beam of

[a] Born, *Physikal. Z.* **21**, 578, 1920.
[b] Bielz, *Z. Physik*, **32**, 81, 1925.

circular cross section, enables the mean free path in virtue of (2·12) to be evaluated from the equation

$$\lambda = \frac{l_2 - l_1}{\log\left(\dfrac{M_{l_2}^o}{M_{l_1}^o}\cdot\dfrac{M_{l_1}^p}{M_{l_2}^p}\right)} \qquad \ldots\ldots(2\cdot13).$$

Values of the order of magnitude demanded by theory were obtained.

When a detector which measures the beam intensity quantitatively is available, the mean free path is obtained from the single equation

$$I_p = I_o e^{-l/\lambda_p} \qquad \ldots\ldots(2\cdot14),$$

where I_o, I_p are the intensities measured at a point behind the image slit distant l cm. from the source, with a perfect vacuum and with a pressure p in the collimator chamber respectively. In case the substance under investigation can be completely removed from the collimator chamber by condensation at the walls, I_o and I_p can be determined successively at the same source pressure.

Measurements of the mean free path of potassium in nitrogen, using the surface ionisation detector, have been reported by Weigle and Pleasant.[a] They obtained values for λ which are undoubtedly too large. From their account, it appears that they did not have an image slit. If this was the case, large λ values would in fact be measured, because potassium atoms *scattered* by the nitrogen in appropriate directions would arrive at the wire and be counted as beam atoms. Hence the weakening of the direct beam would be spuriously small, and consequently λ spuriously large.

With non-condensable gases it is not possible to measure I_o directly (see Chapter 1, p. 21). We assign therefore to I_0 in (2·14) a different meaning: namely, the intensity which the beam would have if it were not, as is actually the case,

[a] Weigle and Pleasant, *Phys. Rev.* 36, 373, 1930. (Abs.)

weakened through scattering by the alien molecules present in the collimator chamber. Now clearly I_o is directly proportional to the quantity of gas issuing from the source slit per second; but so also is the pressure p in the collimator chamber, if a constant pump speed is assumed. We can therefore set $I_o = c.p$. On the other hand, λ is inversely proportional to p; that is $\lambda_p = \lambda_o/p$, where if p is measured say in millimetres of mercury, λ_o is the mean free path at a pressure of 1 mm. I_p can therefore be expressed as a function of the pressure p; thus if l is the distance between source slit and image slit

$$I_p = c.p.e^{-p.l/\lambda_o} \qquad \ldots\ldots(2\cdot15),$$

it being assumed that the pressure in the observation chamber

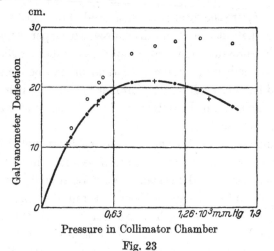

Pressure in Collimator Chamber

Fig. 23

is negligibly small. I_p is a maximum for that value of p which makes $p.l/\lambda_o = 1$. To make a measurement, the intensity is plotted as a function of the pressure in the collimator chamber, and the value of the latter at the maximum intensity is observed. Then $\lambda_p = \lambda_o/p = l$.

The determination of λ for hydrogen has been carried out by Knauer and Stern,[a] using the apparatus of Fig. 13, with

[a] Knauer and Stern, Z. Physik, 53, 766, 1929 (U. z. M. 10).

a hot wire manometer as detector. In Fig. 23, taken from their paper, the circles are the measured values of I_p, as determined from the galvanometer deflections. These values have to be corrected for the effect of the image slit acting as source slit for the collimator chamber. The intensity arising from this cause was for each point measured outside the beam and subtracted from the observed I_p, giving the corrected values represented by the dots. The maximum I_p lies at $p = 1 \times 10^{-3}$ mm.; l was 4 cm.; hence $\lambda_p = 4$ cm. The crosses in Fig. 23 give the values of I_p calculated from (2·15) with λ_0 set equal to 4×10^{-3} cm. The excellent agreement of the values so calculated with observation justifies the assumptions made in deriving (2·15).

The value of the mean free path of hydrogen found by Knauer and Stern is only about 0·44 times that derived by the standard viscosity or conductivity methods. The reason is that the standard methods require an intimate encounter in order that the molecules may exchange energy and momentum in amounts capable of affecting the viscosity or heat conductivity of the gas. The molecular ray method on the other hand counts as a collision an approach of two molecules sufficiently close to deflect them very slightly out of their paths; with narrow slits angular deflections of less than 10^{-4} are detectable. A wide field is here opened up in the study of the scattering of molecular rays by gases, for a knowledge of the dependence of the number of molecules scattered on the angle of scattering would throw much light on the mechanism of the molecular encounters, and on the energy and momentum of the normal state. Preliminary experiments in Stern's laboratory, and latterly in the Laboratory of Physical Chemistry at Cambridge, have already demonstrated the essential applicability of the method.

Chapter 3

PROBLEMS OF THE GAS-SOLID INTERFACE

The kinetic theory of gases, as developed at the hands of Clausius, Maxwell and Boltzmann, was concerned chiefly with gases at pressures such that the mean free path was small compared with the dimensions of the containing vessel. The intermolecular collisions were thus of much greater import-ance in the theory than the collisions of the molecules with the walls of the vessel. During the last twenty years however it has become possible to produce with ease pressures so low that the mean free path of the gas molecules is many hundreds of times greater than the dimensions of any ordinary contain-ing vessel, and attention was thus inevitably directed to the processes occurring when molecules impinge on a solid surface.

Broadly speaking, either one of two things according to the conditions may happen when a molecule strikes a solid sur-face. Either it returns instantaneously to the gas phase, or it is condensed on the surface. In the latter case, it may remain on the surface for a short but finite time only, and then return to the gas phase; or it may remain permanently on the surface as a unit in the formation of a condensate.

The molecules which are restituted from the surface may therefore have had one or other of two previous histories: either they have returned instantaneously to the gas phase on impact, or they have lingered on the surface for a finite time before leaving it. In the first case, the molecules have been *reflected* from the surface, in directions which have a de-finite probability relation to the direction of incidence. In the second case, they have been *scattered* from the surface in direc-tions which, since the restitution is governed entirely by the conditions on the surface, can bear no relation to the direction of incidence. The possibility of scattering and reflection occurring simultaneously at a surface is of course not excluded.

If the surface is macroscopically rough, the molecules will emerge from its hills and valleys in random directions, even though each one has been reflected on impact. Thus unless the surface is smooth, it is impossible to say *a priori* whether reflection or scattering or both simultaneously has occurred at it.

It is to the investigation of scattering and reflection that the molecular ray method finds one of its most striking applications. The ability to confine the directions of incidence of the impinging molecules within definitely known limits makes the determination of the directions of emergence a measurement of the utmost possible directness and precision; and indeed, results have been obtained from a study of the restitution of molecular rays from solid surfaces which could not conceivably have been achieved otherwise.

SCATTERING

Suppose that a molecular beam is incident in a given direction on a surface: What laws govern the directions of the restituted molecules? Let us imagine to begin with that the reflection coefficient ρ of the surface is zero; then all the molecules which are restituted from the surface have lingered for a finite time upon it, and their directions of emergence are governed entirely by the conditions obtaining on the surface. The surface is therefore analogous to a black body in optics, and the law governing the distribution of directions of the scattered molecules is the analogue of the Lambert law of reflection. Translated into molecular terms, this becomes the Knudsen Law of Molecular Scattering,[a] namely:

If a group of n molecules, whose directions lie within an element of solid angle $d\omega'$, strike a surface for which $\rho = 0$,

[a] Knudsen, *Ann. Physik*, **48**, 1113, 1915.

then the fraction dn of those scattered from the surface within an element of solid angle $d\omega$, making an angle ϕ with the normal to the surface, is

$$dn = 1/\pi . n . \cos \phi . d\omega \qquad \ldots\ldots(3\cdot1).[a]$$

The Knudsen law has been established experimentally in a number of molecular ray experiments. A first rough test of the law was made by Wood,[b] who allowed a mercury beam to fall on a plane glass target placed at the centre of a bulb. The atoms restituted from the target were condensed on the bulb, and the deposit was found to be densest around the point of intersection of the perpendicular from the target with the bulb, falling off gradually in all azimuths with increasing distance from that point. He continued the experiments with cadmium,[c] which gives a deposit permanent at room temperature. The deposit thickness was estimated by observing photometrically the transmission of red light by the film. The Knudsen law was obeyed within the probable error (not stated) of the experiment.

Knudsen[d] avoided the difficulty of a point to point measurement of a non-uniform deposit, and in general the uncertainty of photometric estimation of deposit thickness, by making the scattering surface itself part of the receiving sphere. In that case it is easy to show that the deposit is *uniform* if the restitution follows the Knudsen law. Knudsen's

---

[a] I have preferred to call the law as enunciated the Knudsen Law rather than the current term "Cosine Law". The term cosine law properly refers to the restitution of molecules from a surface in temperature equilibrium with a gas above it. In that case, a cosine law of restitution is demanded by the Second Law of Thermodynamics, as has been pointed out by Gaede (*Ann. Physik*, 41, 331, 1913), Epstein (*Phys. Rev.* 23, 710, 1924), and recently by Clausing (*Ann. Physik*, 4, 533, 1930). The cosine law is valid for the most general kind of restitution (scattering + reflection), or for either particular type when present to the exclusion of the other. The Knudsen law is applicable to scattering alone.

[b] Wood, *Phil. Mag.* 30, 300, 1915.

[c] *Ibid.* 32, 364, 1916.

[d] Knudsen, *Ann. Physik*, 48, 1113, 1915.

arrangement is shown schematically in Fig. 24. The substance used was mercury. Samples taken from three separate regions of the deposit on the cooled portion of the bulb showed on weighing perfect uniformity.

It has already been pointed out that a macroscopically rough surface will simulate an ideal surface with $\rho = 0$, even though ρ may actually have a positive value. There has always been some doubt as to whether the surfaces used by Wood and by Knudsen were not

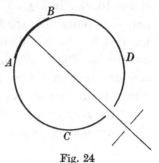

Fig. 24

AB heated; $ACDB$ cooled.

indeed macroscopically rough. Recently Taylor[a] has made extremely accurate observations on the restitution of Li, K and Cs beams from *cleavage planes* of NaCl and LiF under

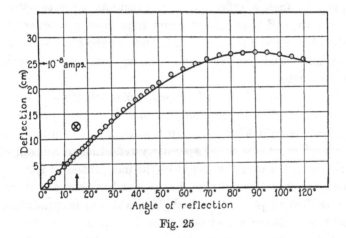

Fig. 25

conditions which, as we shall see, definitely excluded the possibility of simulated scattering. The beams, formed in the usual way, impinged on the crystal face at glancing angles of from 2° to 60°. The intensity distribution of the restituted

[a] Taylor, *Phys. Rev.* **35**, 375, 1930.

molecules was explored with a surface ionisation detector. It was found that the Knudsen law was obeyed with an accuracy of 1 in 10,000. In Fig. 25, taken from Taylor's paper, the solid line represents the angular distribution of scattering demanded by the Knudsen law; the circles are the observed intensities. The results finely demonstrate the correctness of the idea that molecules may linger on a surface, and then return to the gas phase; for only by some such mechanism could the Knudsen law conceivably have been followed under the conditions of Taylor's experiments.

REFLECTION

The simplest possible type of reflection of molecules at a surface is *specular* reflection; that is, the incident beam, the reflected beam and the normal to the surface at the point of impact all lie in the same plane; and the angle of incidence equals the angle of reflection: all in exact analogy with optics. In the case of a molecular beam it is most convenient to characterise the direction of incidence by the *glancing angle*, that is the angle between the impinging beam and the plane of the surface.

The *primary* condition for the specular reflection of molecules at a surface is very simply found if account is taken of the wave nature of matter. It is a familiar fact in optics that any surface can be made specularly reflecting to a beam of light provided that the height of the inequalities of the surface projected on the beam direction is less than one wavelength. Thus a piece of smoked glass, which at perpendicular incidence will not reflect light from a lamp, gives a clear reflected image at near grazing incidence, the colour of the image being at first red, becoming white as the glancing angle is decreased. If θ_0 is the glancing angle, the condition for specular reflection is

$$h \sin \theta_0 \sim \lambda \qquad \ldots\ldots(3\cdot2),$$

where h is the average height of the inequalities on the surface.

Now the inequalities on the most perfect possible mechanically polished surface are of the order 10^{-5} cm. The de Broglie wavelength for hydrogen at room temperature is of the order 10^{-8} cm. Inserting these values in (3·2) we see that specular reflection of molecular beams at mechanically polished surfaces is only possible at very small glancing angles, namely $\theta_0 \sim 10^{-3}$, or a few minutes of arc.

Knauer and Stern[a] have found that specular reflection of a hydrogen beam at room temperature from very carefully polished speculum metal does in fact occur in the expected range of glancing angle, as may be seen from the following table taken from their paper:

Glancing angle:	1.10^{-3}	$1\frac{1}{2}.10^{-3}$	2.10^{-3}	$2\frac{1}{4}.10^{-3}$
Reflecting power:	5	3	$1\frac{1}{2}$	$\frac{3}{4}$ per cent.

They found further that the reflecting power at a given glancing angle was increased about one and a half times by cooling the source to $-150°$ C. This is in complete agreement with the theory, for the de Broglie wavelength is thereby increased in the ratio $\sqrt{300/130} = 1·5_2$ (see (4·5)).

The most perfect surface in nature is the cleaved surface of a crystal. Even the surface of a single crystal is however not absolutely smooth, because the ions in the surface are in temperature vibration about their positions of equilibrium. Now the amplitude of the temperature oscillations is of the order of an Ångstrom (10^{-8} cm.); and following (3·2) we should expect the specular reflection of for example a helium beam at room temperature (most probable de Broglie wavelength $0·57 \times 10^{-8}$ cm.) to become marked at glancing angles less than twenty to thirty degrees.

The curves obtained by Estermann and Stern[b] show precisely the expected form (Fig. 26). In the figure the reflecting power of LiF for a helium beam at two different temperatures

[a] Knauer and Stern, Z. *Physik*, **53**, 779, 1929 (U. z. M. 11).

[b] Estermann and Stern, Z. *Physik*, **61**, 114, 1930 (U. z. M. 15).

has been plotted against the glancing angle. The reflecting power is taken to be the ratio of the maximum intensity of the reflected beam to the maximum intensity of the parent beam, both intensities being expressed as the galvanometer deflec-

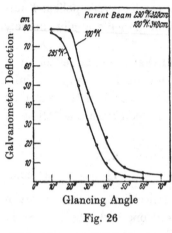

tions obtained with a hot wire gauge detector. It will be noticed, first, that the reflecting power at a given angle is greater for the beam at 100° K., which has the longer de Broglie wavelength; second, that the reflecting power begins to fall off at smaller glancing angles for the beam at 295° K., with the shorter de Broglie wavelength. Helium incident at NaCl, and H_2 at LiF and NaCl show the same behaviour.

Fig. 26

Johnson[a] has obtained quite similar results for the specular reflection of atomic hydrogen from rocksalt.

Apart from any working hypothesis, these experiments have shown that the coefficient of specular reflection ρ^* is a function both of the velocity of the impinging molecules and of the glancing angle. We can write

$$\rho^* = f(v, \theta_0) \qquad \ldots\ldots(3\cdot3),$$

which is a very important result, inasmuch as it has been customary to assume ρ^* independent of v and θ_0.

The hypothesis that the crystal acts as a matt surface to the incident de Broglie waves because the ions in the surface are in temperature vibration receives very striking confirmation from the following fact of observation: For a beam of given de Broglie wavelength incident on a crystal surface at a given glancing angle, the reflecting power is increased by

[a] Johnson, *J. Franklin Inst.* **206**, 308, 1928; *ibid.* **207**, 635, 1929.

lowering the temperature *of the crystal*. The effect has been observed by Knauer and Stern,[a] Estermann and Stern,[b] and Johnson.[c] It is probable that the method could be made capable of giving accurate information about the temperature motion of the ions in the surface lattice; and Stern has indeed suggested that it may be possible, by extrapolating the observations to the absolute zero, to obtain new evidence for the existence of the zero point energy postulated by quantum theory.

We may then feel confident that the necessary conditions for specular reflection are now fairly well understood. It must be emphasised however that condition (3·2), while being *necessary* is not *sufficient*; for clearly if the molecules are restituted from the surface after having lingered upon it for a finite time, specular reflection cannot be obtained at any glancing angle no matter how smooth the surface may be. This is strikingly demonstrated by the experiments of Taylor (see above, p. 81). At the smaller glancing angles (3·2) was undoubtedly fulfilled, yet Taylor failed to observe the slightest trace of specular reflection, although one-hundredth of 1 per cent. could have been detected—the cross in Fig. 25 indicates the galvanometer deflection which would have been obtained at $\theta = \theta_0 = 15°$ if one-tenth of a per cent. of the incident atoms had been specularly reflected.

The factors which decide whether a molecule shall be instantaneously reflected, or adsorbed and re-evaporated from a surface are still obscure. The results so far obtained seem to show, however, that the critical data of the impinging gas or vapour are of predominant importance for reflection. Thus Knauer and Stern[a] found that of the gases H_2, He, Ne, Ar, CO_2, only H_2 and He gave strong specular reflection at NaCl;

[a] Knauer and Stern, *Z. Physik*, **53**, 786, 1929 (U. z. M. 11).

[b] Estermann and Stern, *Z. Physik*, **61**, 115, 1930 (U. z. M. 15).

[c] Johnson, *J. Franklin Inst.* **207**, 635, 1929; *ibid.* **210**, 145, 1930.

[d] Knauer and Stern, *Z. Physik*, **53**, 785, 1929 (U. z. M. 11).

and of these He more than H_2. It might at first sight appear surprising that, as Johnson's experiments show, atomic hydrogen is so strongly reflected. However, Eisenschitz and London[a] have found theoretically that the critical data for H lie between those of H_2 and He; if their results are accepted H falls naturally into line with the other gases.[b]

ADSORPTION AND CONDENSATION

The Time of Adsorption. The idea that the true scattering of molecules at a surface (as distinct from what we have called simulated scattering) arises from the adsorption and subsequent re-evaporation of the molecules at the surface receives its strongest support from the fact that the existence of a finite time of adsorption[c] ("Verweilzeit", "time of lingering") can actually be demonstrated experimentally.

If ν molecules are incident per cm.^{2} per second on a surface, and if n molecules are at any instant adsorbed on unit area of the surface, then it can be shown that for equilibrium

$$n = \tau'\nu \qquad \ldots\ldots(3\cdot4),$$

where τ' is the mean life of an impinging molecule on the surface.[d] Knowing the area of surface available for adsorption and the amount of gas adsorbed, n can be evaluated; ν can be calculated from the equilibrium pressure of the gas.

In this way Wertenstein[e] finds τ' for mercury on glass at room temperature to be of the order 10^{-5} sec. τ' for a number of gases on mica and glass can be computed from data given

[a] Eisenschitz and London, *Z. Physik*, **60**, 516, 1930.

[b] The specular reflection of Zn at rocksalt reported by Zahl (*Phys. Rev.* **36**, 893, 1930) and of Cd and Hg by Ellett and Olson (*ibid.* **31**, 643, 1928) is difficult to account for theoretically. It is conceivably a purely classical effect.

[c] The term "time of adsorption" is due to Clausing, *Ann. Physik*, v, **7**, 492, 1930.

[d] Cf. Clausing, *Ann. Physik*, v, **7**, 489, 1930, for a neat derivation of this equation.

[e] Wertenstein, *J. de Physique*, **4**, 281, 1923.

by Langmuir[a] to be of the order 10^{-5} sec. for a surface temperature of 90° K.[b] The weakness of the method lies of course in the fact that the area of surface available for adsorption by no means necessarily corresponds to the geometrical area of the surface.[c] The values found by Langmuir and by Wertenstein can therefore only be regarded as giving the order of the effect, in spite of the fact that they are customarily quoted to three places of decimals.

It should be observed that the time τ' in (3·4) is *not* equal to the time of adsorption unless the reflection coefficient ρ of the surface is zero. Thus if ν molecules per cm.^{2} per sec. impinge on a surface for which the reflection coefficient (for that species of molecule) is ρ, then a number $(1 - \rho)\, \nu$ are adsorbed, and remain on the surface on the average τ seconds. On the other hand, a number $\rho\nu$ are reflected instantaneously from the surface.[d] Hence

$$\tau' = \frac{(1 - \rho)\, \nu\tau + \rho\nu.0}{\nu} = (1 - \rho)\, \tau \quad \ldots\ldots(3·5),$$

where τ is the mean time of adsorption *reckoned from the moment of impact*. Thus only in case $\rho = 0$ is the mean life τ' *of an impinging molecule* on the surface identical with the true time of adsorption τ, as defined above.[e]

An ingenious molecular ray device, which ideally should be capable of measuring both τ' and τ, was suggested by Holst, and has been brought to some degree of practical realisation by Clausing.[f] The principle of the method will be clear from an examination of Fig. 27. A is an oven from which cadmium

[a] Langmuir, *J. Amer. Chem. Soc.* 40, 1390, 1918.

[b] See Hückel, *Adsorption und Kapillarkondensation*, Leipzig, 1928, p. 173.

[c] Cf. Bowden, *Nature*, 122, 647, 1928.

[d] It is assumed for simplicity that ρ is independent of the direction of incidence of the molecules. This is however not in general true: see (3·3) above.

[e] See for example Hückel, *loc. cit.* p. 170; also Clausing, *Ann. Physik*, 7, 489, 1930.

[f] Clausing, *Physica*, 8, 289, 1928.

is vaporised; D is the circular image aperture defining a beam, which impinges on the disc E. The disc is held at a temperature T sufficiently high to restitute the impinging atoms to the cold body C, where they are condensed. With the disc at rest, the atoms are deposited on C symmetrically around the lip of the aperture D. Suppose now that the disc is set in rapid rotation about the axis F. Then an atom which is adsorbed on the disc for an arbitrary time τ_a, reckoned from the instant of impact, will be carried round on the rotating disc for a certain distance before it is restituted to the body C, which it will strike at a distance from the aperture

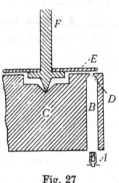

Fig. 27

$$\Delta_a = 2\pi r n \tau_a,$$

where r is the radius of E and n the number of revolutions per second. A little consideration will show that the *centre of gravity* of the deposit on C is displaced in the direction of motion of the disc by an amount

$$\bar{\Delta} = 2\pi r n \tau'.$$

It is necessary in practice to correct for the fact that the rotating disc contributes momentum to the molecules leaving it, the displacement due to this cause being

$$\delta = \frac{2\pi r n}{u} \cdot d,$$

where u is the velocity, assumed to be that corresponding to the temperature T of the disc, with which the atoms leave E, and d is the distance between E and C. Thus the observed displacement of the centre of gravity of the deposit $\Delta = \bar{\Delta} + \delta$, and

$$\tau' = \frac{\Delta - \delta}{2\pi r n} \qquad \ldots\ldots(3\cdot6).$$

As a numerical example, we may take $n = 200$, $r = 2$ cm.,

$\Delta - \delta = 10^{-3}$ cm., the smallest displacement readily measurable; then $\tau' = 0.4 \times 10^{-6}$ sec.

In Clausing's experiments the momentum correction δ was so large as to mask $\overline{\Delta}$ completely. The observed displacement Δ agreed quite closely with the calculated displacement δ; in fact, it was certain that $\Delta - \delta < 10^{-3}$ cm. Hence $\tau' < 10^{-6}$ sec. On the other hand, the fact that $\Delta \cong \delta$ showed that the atoms had acquired almost completely the velocity of the rotating disc, and this can only be understood if τ' is great in comparison with the period of atomic vibration. Therefore Clausing could conclude that

$$10^{-12} \text{ sec.} < \tau' < 10^{-6} \text{ sec.}$$

for cadmium on glass, picein, mica and copper at about 200° K.

The method would probably have yielded a more definite result if it had been used with a substance, and with a beam intensity, for which the transition temperature of condensation (q.v.) is very low; for τ' depends on the temperature of the surface according to the relation

$$\tau' = (1 - \rho)\, \tau = \frac{C}{\sqrt{T}} . e^{u_0/RT} \qquad \ldots\ldots(3.7),$$

where u_0 is the heat of adsorption per mole, and C is a constant.[a] Thus if the temperature T of the rotating disc in Fig. 27 is very low, τ' is correspondingly great, and the correction term δ is of less relative importance.

In the ideal case of negligibly small δ, measurement of the intensity distribution in the deposit would serve to differentiate between the atoms which are instantaneously reflected from the disc ($\Delta_a = 0$) and those which are adsorbed for a finite time thereon ($\Delta_a > 0$). Thus both τ' and τ would be measurable. It seems doubtful however if the exceedingly difficult technique of the Clausing device could be so far refined as to realise the ideal conditions.

[a] See for example Hückel, *loc. cit.* p. 153.

Lateral Motion. There is good evidence that during the time of adsorption the molecules do not necessarily remain in one position, but may move about at random on the surface. The evidence has been obtained chiefly from a study of the formation of permanent *deposits*.

Thus if a silver deposit from a molecular beam, the intensity of which is sensibly constant over its cross section, is laid down on glass to a calculated thickness of 0·1 to 2 atoms thick, an ultramicroscopic examination of the deposit reveals the appearance seen in Plate II, Fig. 28. Here then is no question of a uniform deposit: the silver atoms, which *arrived* at a uniform rate over the whole surface, have gathered into isolated nuclei, each of which must consist of at least 1000 atoms.[a]

Cockcroft[b] obtained brilliantly coloured metal films, deposited on copper from a molecular stream effusing from a circular oven aperture, of calculated thickness 2–3 atoms. The colours of such very thin films cannot arise from interference, and must be identified with the so-called resonance colours described by Wood and by Maxwell Garnett,[c] which are the property of an aggregate of very small isolated metallic nuclei. Thus the appearance of these coloured films is in complete agreement with the ultramicroscopic structure found by Estermann.

Cockcroft obtained other evidence for the existence of lateral motion, by placing a narrow wire across the stream. Plate II, Fig. 29 is a photograph of a portion of the deposit, and shows very clearly how the atoms have crept in under the shadow of the wire and have there formed condensation nuclei.

The existence of lateral motion is very prettily shown in

[a] Estermann, *Z. physikal. Chem.* 106, 403, 1923.

[b] Cockcroft, *Proc. Roy. Soc.* 119, 293, 1928.

[c] Maxwell Garnett, *Phil. Trans.* 203, 385, 1904; see also Wood's *Physical Optics*, 2nd ed. 1923, pp. 635–6, 643–7.

Plate II

Fig. 28

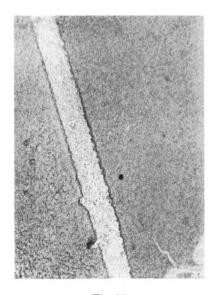

Fig. 29

Lateral Motion

some further experiments of Estermann.[a] A wedge-shaped stencil was placed over a target held at a temperature such that the formation of a deposit from a cadmium beam was just possible. The intensity of the beam was uniform over the whole length of the wedge. Nevertheless the deposit appeared first at the thick end of the wedge, and only much later at the tip. Now it is clear that a deposit can form only so long as the number of atoms impinging on a given area of surface is greater than the number leaving that area, either through re-evaporation or by lateral motion. In Estermann's experiments the rate of escape of the atoms from the deposit on account of re-evaporation was uniform over the length of the wedge. On the other hand, the atoms could escape from the bombarded area in virtue of their lateral motion more readily at the narrower parts of the deposit; consequently the time of appearance of the deposit increases in going from the base to the apex of the wedge.[b]

The Two-dimensional Gas Phase. The layer of adsorbed molecules in random lateral motion present on a solid surface which is scattering the molecules impinging on it from the gas phase, may be formally represented as a two-dimensional gas. We may then with Volmer[c] set up for it an equation of state analogous to that of van der Waals for a three-dimensional gas, namely

$$\left(\pi + \frac{\alpha}{\omega^2}\right)(\omega - \beta) = RT \qquad \ldots\ldots(3\cdot8),$$

in which π is the two-dimensional pressure, that is, the force per unit length of boundary. α and β are equivalent to van der Waals' a and b respectively; β is however taken to

[a] Estermann, *Z. Physik*, **33**, 320, 1925; see also Gerlach, *Ergebnisse der exakt. Naturwiss.* III, 1924, p. 186.

[b] Evidence for the lateral motion of molecules over solid surfaces other than that obtained with the molecular ray method exists. In this connection see Volmer and Estermann, *Z. Physik*, **7**, 1; 13, 1921, and Volmer and Adhikari, *ibid.* **35**, 170, 1925; *Z. physikal. Chem.* **119**, 46, 1926.

[c] Volmer, *Z. physikal. Chem.* **115**, 253, 1925.

be twice the molar area, instead of four times the molar volume as in three dimensions. ω is the area of the surface on which is adsorbed one mole. T is the temperature of the surface.

We are led thus to postulate the existence of a *critical temperature $T_{c \cdot}$* for the two-dimensional gas, below which it is possible for a two-dimensional liquid to coexist with its saturated two-dimensional vapour on the surface. If we call the two-dimensional saturation pressure π', we have in analogy with the Clapeyron equation

$$\pi' = k_1 . e^{-\Delta u/RT'} \qquad \ldots\ldots(3\cdot9),$$

where Δu is the latent heat of vaporisation of the two-dimensional liquid, and k_1 is a constant; the surface temperature T'' being subject to the limitation $T'' < T_{c \cdot}$.

Now assuming for simplicity a homogeneous surface, it can be shown by means of a quite general statistical argument that the molar density ρ (moles per cm.³) of a three-dimensional gas in equilibrium with the adsorbed layer is connected with the density σ (moles per cm.²) of the two-dimensional *gas*, in the limit for small σ, by the relation

$$\rho = k_2 . e^{-u_0/RT} . \sigma \qquad \ldots\ldots(3\cdot10),$$

where k_2 is a constant, and u_0 the heat of adsorption per mole.[a] Or for a given temperature we can write

$$p = k_3 . e^{-u_0/RT} . \pi \qquad \ldots\ldots(3\cdot11),$$

where p is the pressure of the three-dimensional gas and k_3 is a constant.

Combining (3·9) and (3·11), we have the result that to every two-dimensional saturation pressure π' corresponds an external pressure p' such that

$$p' = a . e^{-(u_0+\Delta u)/RT'} = a . e^{-u/RT'} \qquad \ldots\ldots(3\cdot12),$$

where a is a constant. Thus to a given surface temperature

[a] Cf. Hückel, *loc. cit.* pp. 157 ff.

T' ($T' \gg T_{c^*}$) corresponds a definite external pressure p' (or beam intensity I') at which the two-dimensional liquid phase is formed.

Transition Phenomena.[a] It has been found that to a given target temperature corresponds a certain minimum intensity of a beam of *metallic atoms* below which a deposit does not form on the target, no matter how long it is exposed to the beam. Conversely, for a beam of given intensity there exists a definite temperature of the target above which the formation of a deposit is not possible.

Knudsen[b] and Wood[c] established the existence of a *transition temperature* of the target for a variety of metals. They failed however to observe the existence of a corresponding *transition intensity*, and regarded the transition temperature therefore as a constant characteristic of a given target and beam species.

The dependence of the transition temperature on the beam intensity was demonstrated in an elegant experiment of Chariton and Semenoff,[d] who examined the deposit formed on a plate, along the length of which a constant temperature gradient was maintained, by the evaporation of cadmium from an electrolytically coated platinum wire stretched above it. Fig. 30 is a sketch of the arrangement as used by Cockcroft;[e] the curved shape of the upper boundary shows clearly that the transition temperature T' is not a characteristic constant, but depends on the stream intensity I.[f]

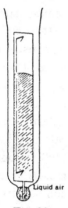

Fig. 30

[a] See Footnote, p. 27.   [b] Knudsen, *Ann. Physik*, **50**, 472, 1916.

[c] Wood, *Phil. Mag.* **32**, 364, 1916.

[d] Chariton and Semenoff, *Z. Physik*, **25**, 287, 1924.

[e] Cockcroft, *Proc. Roy. Soc.* **119**, 295, 1928.

[f] The fact that I must be greater than I' for condensation to take place would account for the fact that Henderson (*Proc. Camb. Phil. Soc.* **25**, 344, 1929) failed to obtain a geometrically defined deposit from his extremely weak beams of polonium.

The first quantitative measurements were made by Estermann[a] with beams of cadmium and mercury. The equivalent pressure p' was calculated from the transition intensity I' (in molecules/cm.^{2} sec.) from the relation $p' = \sqrt{2\pi k m T'} \cdot I'$, where k is Boltzmann's constant and m the mass of an atom. It was found that log p' plotted against $1/T'$ gave a straight line, in agreement with (3·12).

Now it emerges from these results that p' is far greater than the saturation pressure p_s above the solid metal. Thus for example p' for cadmium incident at a copper target held at ca. 200° K. is of the order 10^{-7} mm.; whereas p_s for massive cadmium at the same temperature is found by extrapolation of Egerton's values[b] to be very roughly 10^{-19} mm. Thus for metals it appears to be generally true that the equilibrium of the two-dimensional liquid when it is once formed is unstable; and a true *condensation* occurs below the transition temperature.

It is unlikely that similar transition phenomena will be demonstrable with gases, except at extremely low temperatures of the surface. It must be remembered that the critical temperature $T_{c^{\bullet}}$ of the two-dimensional gas lies well below that of T_c of the corresponding three-dimensional gas; for the external fields of the adsorbed molecules are partially saturated by those of the surface, and the cohesion constant a in (3·8) is consequently small. The phenomena can occur with metals or easily condensable vapours because a (and T_c) is comparatively large.

Surface Conditions and Transition Phenomena. The values of u in (3·12) found for the same beam species by different observers show very bad agreement. Thus Estermann found for cadmium incident on copper $u_{Cd, Cu} \cong 3000$ cal./mole; Cockcroft, whose observations extended over a far wider

[a] Estermann, *Z. f. Elektrochem.* **31**, 441, 1925.
[b] Egerton, *Phil. Mag.* **33**, 33, 1917.

range of beam intensity and surface temperature, obtained $u_{Cd, Cu} = 5680$ cal./mole.

Now the surfaces used by Estermann and by Cockcroft were certainly covered to a greater or less extent by a film of gas, or vapour from tap grease. Hence reproducibility of the surface conditions is under these circumstances not to be expected. It is extremely probable, moreover, that the whole transition phenomenon as obtained with alleged metal surfaces depends on the presence of an adsorbed gas or vapour layer on the surface of the metal. It must be remembered that the whole argument of the preceding articles is based on the tacit assumption that the forces between the adsorbed molecules are great compared with those between the adsorbed molecules and the surface; for otherwise the cohesion constant α in (3·8) would be negligibly small, and the critical temperature T_{c^*} extremely low. It would follow that transition phenomena should occur with metals only on surfaces that are non-polar, or weakly polar. We have really no right to expect their occurrence on an outgassed metal surface, but rather a deposition of the metal at even the weakest intensities, not in isolated nuclei, but in regular layers of restricted mobility which follow the crystal lattice of the underlying surface. Such metal deposits on outgassed metal surfaces have been much studied in the field of Thermionics, and have in fact the properties described.

An experiment of Cockcroft[a] shows that the effect of a comparatively uncontaminated metallic surface is in the expected direction. A cooled copper plate was placed opposite the circular aperture of an oven charged with cadmium, and the cadmium was evaporated at a temperature such that the stream density was well below the transition value for the temperature of the receiving surface. If now from a second oven placed alongside the cadmium oven silver was evaporated on to the receiver, cadmium immediately deposited on

[a] Cockcroft, *Proc. Roy. Soc.* **119**, 306, 1928.

the freshly formed silver surface, the centre of the deposit being opposite the aperture *of the silver oven.*

Clearly much more work remains to be done towards a complete understanding of the phenomena of condensation at solid surfaces. The molecular ray work which has been accomplished in this field has been mainly qualitative in character: the vacuum technique has been too primitive, and the surfaces in consequence have not been reproducible. The difficulties which would occur in the construction of a molecular ray apparatus which it is required to outgas could not be other than rather formidable, however; and it may even prove impracticable to extend the method to include the study of surfaces under adequate experimental conditions.

Chapter 4

THE DIFFRACTION OF MOLECULAR RAYS

The conditions for the specular reflection of a molecular beam at a solid surface are, as we have seen, intimately connected with the wave nature of the molecules in the beam. We should expect therefore that if a grating were ruled on the surface, *diffraction spectra* of the de Broglie waves would be observable when the conditions for specular reflection at the surface were fulfilled.

Now a mechanically ruled grating could only be used to investigate molecular diffraction near grazing incidence (see p. 83 above). This fact introduces in one sense a notable technical simplification; for, following the analogy of the Thibaud-Compton method of measuring the wavelength of soft X rays, the grating spacing can be made comparatively wide if the grating is used near grazing incidence.[a] Knauer and Stern[b] attempted to detect the diffraction of a hydrogen beam incident at glancing angle ca. 10^{-3} on a grating of some 100 lines per millimetre ruled on speculum metal, but never obtained more than an indication of the first diffraction maximum. The difficulty is mainly one of intensity. If the resolution is to be made sufficiently great to separate the expected first maximum clearly from the directly reflected beam, the intensity of the diffracted beam becomes too small for certain detection. We have seen (Chapter 1) that the maximal intensity of gas beams depends on the square root of the pump speed; hence it is by no means improbable that the measurement of the de Broglie wavelengths of molecules at a ruled

[a] See article by T. H. Osgood, "The Spectroscopy of Soft X Rays", *Reviews of Modern Physics*, **1**, 228, 1929.

[b] Knauer and Stern, *Z. Physik*, **53**, 779, 1929 (U. z. M. 11).

grating may yet be made practicable by the use of extremely fast pumps.[a]

The diffraction of X rays by the regular space lattice formed by the ions in a crystal has long been known, and it seemed at the outset probable that the natural crystal grating could be used to study the diffraction of molecular rays, the de Broglie wavelength for which (10^{-9} to 10^{-8} cm.) is of the same order as the interionic distances in the crystal. There is however an essential difference between the diffraction at a crystal of X rays and electrons on the one hand, and of molecules on the other. The former penetrate in general for a distance of many atomic layers into the crystal, and the grating responsible for the diffraction is the three-dimensional grating formed by the regular arrangement of the ions in space; the molecular rays on the other hand are reflected at the surface of the crystal, and we should therefore expect the appropriate diffraction grating to be the *two-dimensional grating* formed by the regularly spaced ions in the surface. The crystal surface-lattice grating has in fact been used with complete success by Stern and his collaborators[b] to detect and measure the de Broglie waves of H_2 and He, and later by Johnson[c] to study qualitatively the diffraction patterns of atomic hydrogen received at a MoO_3 target.

CROSS-GRATING SPECTRA

A necessary preliminary to the discussion of these experiments is a vivid mental picture of the way in which diffraction spectra are formed at a two-dimensional or cross grating. We begin with the one-dimensional case, from which the two-dimensional case readily follows; and consider in the first instance a simple mechanical analogy. In Fig. 31, consider a

---

[a] The diffraction of *electrons* at a ruled grating has been demonstrated by Rupp (*Z. Physik*, **52**, 8, 1929).

[b] Stern, *Naturwiss.* **17**, 391, 1929; Estermann and Stern, *Z. Physik*, **61**, 95, 1930 (U. z. M. 15). See also Knauer and Stern, *ibid.* **53**, 779, 1929 (U. z. M. 11).

[c] Johnson, *Phys. Rev.* **35**, 1299, 1930; *J. Franklin Inst.* **210**, 135, 1930.

wave incident from above on the particles $a, b, c, \ldots$ etc., which are evenly spaced along the straight line AB. Each particle is set into vibration as the wave passes over it, and sends out from itself as centre secondary spherical wavelets, which are shown in section in the figure. Let us consider the secondary waves, and confine our attention to the plane section of the figure; then it is seen that along certain well-defined directions in that plane the secondary wavelets merge to go forward as a united wave front. Three of these wave-fronts are in-

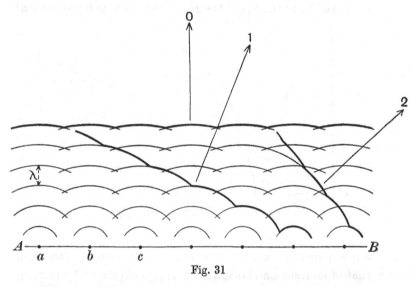

Fig. 31

dicated by the full lines carrying the arrows 0, 1, 2. For all intermediate directions the secondary wavelets do not adjoin crest to crest and hence destroy each other.

The wave fronts advancing in the directions 0, 1, 2, ... n are called diffraction spectra of the zero, first, second, ... nth order, the spectrum of zero order being simply the directly reflected wave. The physical meaning to be attached to the designation "nth order spectrum" will be clear from an examination of the figure. It will be seen that the length of path traversed by a given secondary wavelet from its parent particle to the united wave front differs by a complete number of wavelengths from that traversed by the wavelet arising from the neighbouring particle. For order 0, the distance travelled by each secondary wavelet composing the wave front is the same; for order 1, the distance travelled by a given wavelet is *one* wavelength longer than is the case for its right-hand neighbour; and so on.

Fig. 31 shows the formation of diffraction spectra to the *right* of the zero order only. The configuration is completed by a similar set to the *left* of the zero order, in which the path of

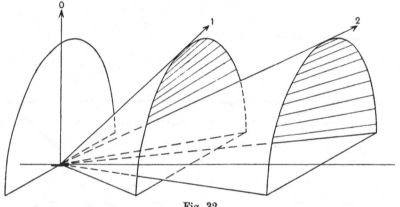

Fig. 32

a given wavelet is a whole number of wavelengths *less* than that of its right-hand neighbour. This is expressed by naming such spectra the $-1, -2, \ldots -n$ order spectra.

One must now recall that the scheme of Fig. 31 is a plane section of the configuration formed by the *spherical* wavelets sent out from the particles $a, b, c, \ldots$ etc. Thus the spectra of order $0, 1, 2, \ldots n$ which are, in the plane of the figure, found along the directions indicated by the arrows, are found in space anywhere on the surface of the cones generated by the rotation of the arrows $0, 1, 2, \ldots n$ about the line AB. The location of the diffraction spectra in space for the zero and first two positive orders is shown in perspective in Fig. 32, in

which can be seen the first and second order cones, together
with the zero order cone, which for the case of normal incidence we are considering degenerates to a disc.

We proceed next to determine the equations to the cones,
and consider in Fig. 33 at once the general case of oblique
incidence. Let a plane wave, wave front mP, be incident at
an angle α_0 on a one-dimensional grating defined by the x-axis
of a Cartesian system of coordinates; and let the first order
diffracted wave, wave front nP', make an angle α with the
x-axis. Then the difference of path traversed by a ray passing

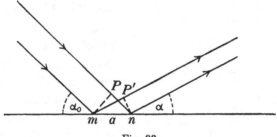

Fig. 33

over particle n, and a ray passing over particle m, namely
$mP' - nP$, must be exactly one wavelength λ. We have

$$mP' - nP = \lambda,$$

or $$a \cos \alpha - a \cos \alpha_0 = \lambda,$$

where a is the distance between adjacent particles. In general
for the nth order spectrum

$$\cos \alpha - \cos \alpha_0 = n . \lambda / a \qquad \ldots\ldots(4\cdot1),$$

which is the equation to a cone of semi-vertical angle

$$\alpha = \cos^{-1} (\cos \alpha_0 + n . \lambda / a)$$

about the x-axis.

Consider next the simplest possible type of two-dimensional
grating (Fig. 34), built up of a family of equidistant one-
dimensional gratings parallel to the axis of x. Evidently, the
grating may equally well be considered as made up of an

identical family parallel to the axis of y. Thus a wave incident on the grating in a direction defined by the direction cosines $\cos \alpha_0$, $\cos \beta_0$, $\cos \gamma_0$ will give rise to two families of diffraction cones, one set as before about the x-axis, the second set about the y-axis, their equations being

$$\left.\begin{array}{l} \cos \alpha - \cos \alpha_0 = h_1 . \lambda/a \\ \cos \beta - \cos \beta_0 = h_2 . \lambda/a \end{array}\right\} \quad \ldots\ldots(4\cdot2),$$

where h_1, h_2 are whole numbers, positive or negative, not necessarily identical.

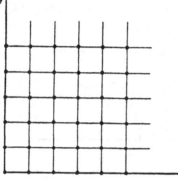

The conditions necessary for the formation of a united wave front are clearly only satisfied for both sets of diffraction spectra simultaneously along the *lines of intersection* of the y-axis cones with the x-axis cones. The diffraction spectra from a two-dimensional grating

Fig. 34

therefore bear two reference numbers, and are designated in general the (h_1, h_2) spectra.

Let us consider in detail (Fig. 35) the simplest possible special case, namely when the incident beam lies in the xz plane. Then from $(4\cdot2)$

$$\beta_0 = 90°, \quad \cos \beta_0 = 0.$$

1. Let $h_1 = h_2 = 0$. We obtain the directly reflected beam, or $(0, 0)$ order, as the line of intersection of the α-cone (cone about the x-axis with semi-vertical angle α, α being in this case equal to α_0) with the β-cone, which here degenerates to the xz plane. The directly reflected beam is indicated by the arrow $(0, 0)$ in the figure.

2. Let $h_1 = 0$, $h_2 = \pm 1$. We obtain now the $(0, \pm 1)$ orders, which are by far the most intense diffraction spectra.

The diffracted beams are given by the lines of intersection of the α-cone ($\alpha = \alpha_0$) and the β-cones about the y-axis with angle β determined by the equation

$$\cos \beta = \pm \lambda/a \qquad \text{......(4.3).}$$

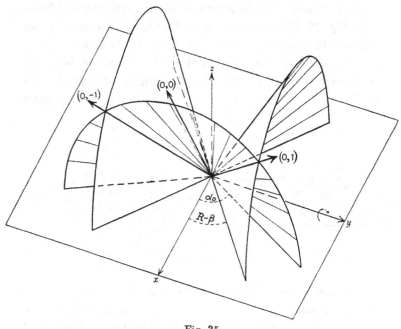

Fig. 35

The $(0, 1)$, $(0, -1)$ orders are indicated by arrows in the figure. It is important to notice that the $(0, \pm 1)$ orders are formed only so long as the angle of diffraction

$$\theta = (90° - \beta) < \alpha_0;$$

otherwise, the α-cone and the β-cones cannot cut.

DIFFRACTION OF MOLECULAR RAYS AT
CRYSTAL SURFACES

It was clearly desirable to make the first attempts to detect the diffraction of molecular rays at the cross grating of a crystal surface, with the crystal oriented to the incident beam as in Fig. 35. The natural ionic surface lattice on the cleavage surface of a crystal differs however from the ideal cross grating of Fig. 34, in that the constituent points are not identical but, reckoning parallel to the cleavage edges, alternately dissimilar (for example, Na·, Cl′; Li·, F′). It was at the outset

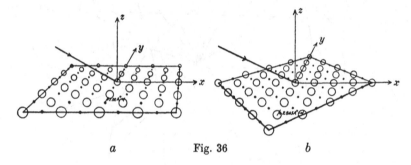

a   Fig. 36   b

by no means clear whether the x-axis of the ionic grating should be taken as coinciding with a row of alternately dissimilar ions, or whether it should be identified with a row of similar ions; that is, whether the crystal should be oriented to a beam incident in the xz plane as in Fig. 36 a or as in Fig. 36 b to obtain the special case of Fig. 35.

The second alternative (Fig. 36 b) was shown to be the correct one by Stern;[a] the experiment yielded at the same time the first direct experimental confirmation of the wave nature of a molecular beam.

A beam of helium, which had the form of a narrow rectangle (5 mm. × 0·5 mm.), was set at an angle of $11\frac{1}{2}°$ "edge on" to

[a] Stern, *Naturwiss.* 17, 391, 1929; see also Estermann and Stern, *Z. Physik*, 61, 95, 1930 (U. z. M. 15).

the crystal (NaCl); that is, the long side of the rectangle lay in the plane of incidence. The reflected beams were received at the slit of a hot wire gauge detector, which was rotatable about an axis perpendicular to the surface of the crystal. With the crystal oriented as in Fig. 36 *b*, diffraction maxima were observed lying symmetrically right and left of the

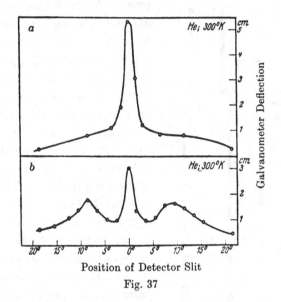

Position of Detector Slit

Fig. 37

directly reflected beam, just as in Fig. 35; on the other hand, no such maxima were found with the orientation of Fig. 36 *a*; nor from the complete theory of the cross grating are they to be expected, if the ionic grating is indeed formed from rows of similar ions. Point to point plots of the diffraction patterns as observed by Stern are reproduced in Fig. 37 *a* (crystal oriented as in Fig. 36 *a*), and Fig. 37 *b* (crystal oriented as in Fig. 36 *b*).

Experiments of Estermann and Stern. In Stern's experiment the $(0, \pm 1)$ diffraction spectra, which as may be seen from Fig. 35 lie nearer to the crystal than the directly reflected beam ($(0, 0)$ order), gained access to the detector on account of the considerable height (5 mm.) of the detector slit. In order therefore to determine correctly the positions of the $(0, \pm 1)$ orders in space, a very short receiving slit is necessary; and ideally, one rotatable not only about the z-axis, but about the y-axis as well.

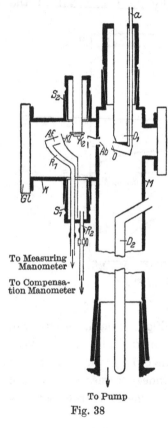

To Measuring Manometer

To Compensation Manometer

To Pump

Fig. 38

It is difficult in practice to provide for both motions of the detector slit, but the necessity for a motion about the y-axis can be avoided if, instead of bringing the detector slit down to meet them, the $(0, \pm 1)$ orders are brought upwards to meet the slit. This is effected by tilting the crystal through a small angle δ about the y-axis in the direction of the arrow in Fig. 35. The angle α_0 is thereby increased, the α-cone opens umbrella-like, and its lines of intersection with the β-cones move up the latter away from the crystal. In this way the $(0, \pm 1)$ orders can by proper choice of δ be directed into the detector slit. The optimum δ, corresponding to maximum intensity received at the detector for a given position of the detector slit, is found by simple trial and error.

The essential features of the apparatus used by Estermann and Stern[a] are seen in Fig. 38. The gas, helium or hydrogen, is led via the tube a to the source slit O, the temperature of which can be lowered by introducing liquid air into the Dewar vessel D_1, or raised by means of a heating spiral (not

[a] Estermann and Stern, *Z. Physik*, **61**, 95, 1930 (U. z. M. 15).

shown). The collimator chamber consists simply of the wide tube M, which sits direct on a large Leybold steel pump. D_2 is a Dewar vessel to remove mercury vapour. The wall of the tube is pierced at Ab by the image slit, which gives access to the observation chamber K containing the crystal K_2 and detector Af. The source and image slits are some 1 mm. long and

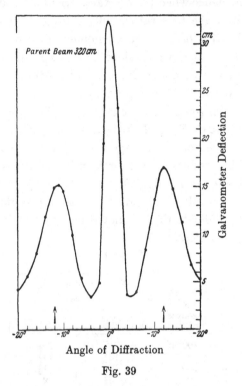

Fig. 39

0·2 mm. wide, the detector slit somewhat longer, 1·5 mm. The crystal grating is oriented with reference to the incident beam by turning the joint S_2. The crystal can also be rotated about an axis perpendicular to the plane of the paper, by a device which is not shown in the figure. The detector is rotatable about an axis perpendicular to the crystal face by turning the joint S_1, which carries at the same time the measuring manometer, the compensation manometer at R_2 (cf. Chapter 1, p. 35), and a shutter Kl.

Estermann and Stern have made a very complete study with this apparatus of the diffraction of molecular rays at crystal surfaces. They investigated not only spectra of order $(0, \pm 1)$, but also those of order $(1, 1)$ which are formed in the plane of incidence when the crystal is oriented as in Fig. 36 a. We shall describe here rather fully a typical set of results; but a realisation of how the phenomena observed follow in the closest detail the behaviour to be expected of a plane wave diffracted at a two-dimensional grating can only be gained by a study of the original paper.

We select for representative discussion the observations on the diffraction of He and H_2 beams incident at a glancing angle α_0 of $18\frac{1}{2}°$ on a cleavage face of LiF, oriented as in Fig. 36 b. The results are set forth in Table IV, No. 3 being shown graphically in Fig. 39, in which every point on the curve has been observed with the optimum angle of tilt δ. The arrows indicate the calculated positions of the maxima of the $(0, \pm 1)$ orders.

Table IV

Gas	Temperature of source	Position of maximum θ'_{max}		No.
		Calc.	Found	
He	100° K.	21°	$15\frac{1}{2}°$	1*
	180	$15\frac{1}{2}$	$14\frac{1}{4}$	2
	290	12	$11\frac{1}{2}$	3
	590	$8\frac{3}{4}$	9	4
H_2	290	17	17	5
	580	12	11	6

* The discrepancy here is merely apparent. See further, p. 110 below.

The maxima are not quite sharp, because the beam is not monochromatic in the de Broglie waves in consequence of the Maxwell distribution of velocities of the molecules in the source. Thus the fraction dI of the total parent intensity I contributed by molecules whose velocities lie between v and $v + dv$ is (equation (2·3))

$$dI = C'.e^{-v^2/a^2}.v^3.dv.$$

Changing the independent variable from v to $v = h/m\lambda$, we have

$$dI = C''.e^{-\lambda_a^2/\lambda^2}.1/\lambda^5.d\lambda = C''f(\lambda)\,d\lambda \ \ \dots\dots(4\cdot4).$$

For the wavelength λ_m of greatest intensity, $\dfrac{dI}{d\lambda} = C''f(\lambda)$ is a

maximum; that is $\dfrac{df(\lambda)}{d\lambda} = 0$, whence $\lambda_m = \sqrt{\tfrac{2}{5}}.\lambda_a$.

Now

$$\lambda_a = h/ma = \frac{30\cdot8 \times 10^{-8}}{\sqrt{TM}}\ \text{cm.,}$$

where T is the temperature of the source and M the molecular weight of the gas. Thus

$$\lambda_m = \frac{19\cdot47 \times 10^{-8}}{\sqrt{TM}}\ \text{cm.} \qquad \dots\dots(4\cdot5).$$

In calculating the position of the diffraction maxima, the fact that the crystal has been tilted through an angle δ about the y-axis must be taken into account. The axis of rotation of the detector slit is then no longer the z-axis, but one (which we may call the z'-axis) which makes an angle of δ with the z-axis. The angle of diffraction θ' which is measured with the crystal tilted is therefore the angle between the plane of incidence and the plane determined by the diffracted beam and the z'-axis. It is easy to show that to a sufficient approximation

$$\tan\theta' = (\tan\theta)\left[1 + \delta\tan\alpha_0\left(1 + \sqrt{1 - \frac{\tan^2\theta}{\tan^2\alpha_0}}\right)\right] \left.\vphantom{\begin{array}{c}\\\\\end{array}}\right\} \ \dots(4\cdot6).$$

whereby $\qquad \tan\theta = \dfrac{\lambda}{a}\cdot\dfrac{1}{\cos(\alpha_0 + \delta)}$

The values of $\theta'_{\max}$ in column 3 of Table IV are obtained by setting $\lambda = \lambda_m$ in (4·6); $a = 2\cdot845 \times 10^{-8}$ cm. (see Fig. 36 b).

The agreement between theory and experiment is clearly very close. It might seem at first sight that No. 1, He at 100° K., is an exception. The discrepancy between the observed value $15\frac{1}{2}°$ and the calculated value 21° of θ'_{max} is however only apparent. It will be observed that for this case, and for this case alone, θ'_{max} (21°) is greater than the glancing angle of incidence $\alpha_0 = (18\frac{1}{2}° + \delta)$. Thus the α-cone and the β-cones for $\lambda = \lambda_m$ do not cut (see Fig. 35). This cannot be corrected beyond a certain point by increasing δ, for the useful angle of tilt is limited. Even for a beam of negligible dimensions, it cannot be made greater than half the incident glancing angle, for otherwise the diffracted beam will be screened from the detector by the crystal surface. For a beam of finite width the limiting value for δ is still smaller. Thus the maximum found at $15\frac{1}{2}°$ is a spurious one, arising from the breakdown of the diffraction patterns for the longer wavelengths.

That the discrepancy really does arise from the fact that the glancing angle of incidence may be too small for the longer wavelengths is clear from Table V. Here are tabulated

Table V

Glancing angle α_0	Temperature of source	Position of maximum θ'_{max}		No.
		Calc.	Found	
$11\frac{1}{2}°$	290° K.	$16\frac{3}{4}°$	$14\frac{1}{4}°$	1
$18\frac{1}{2}°$	290	17	17	2

for comparison the observed and calculated values of θ'_{max} for hydrogen at the same temperature (290° K.) but incident at different glancing angles ($11\frac{1}{2}°$ and $18\frac{1}{2}°$). It is seen that for $\alpha_0 = 18\frac{1}{2}°$, the calculated value of θ'_{max} (17°) is less than the glancing angle, and the agreement between observation and theory is perfect; for $\alpha_0 = 11\frac{1}{2}°$, $\theta'_{max} > \alpha_0$, and the

Plate III

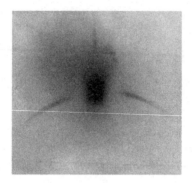

Fig. 40

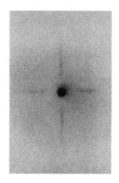

Fig. 41

Diffraction Patterns

spurious maximum at a smaller angle of diffraction at once appears.[a]

Nos. 3 and 6 in Table IV deserve particular attention. The value of θ'_{max} found for helium at 290° K. $(11\frac{1}{2}°)$ is equal within the experimental error to that (11°) found for hydrogen, with half the mass, at twice the absolute temperature (580° K.); in perfect agreement with (4·5). Thus the series of experiments summarised in Table IV constitute a complete quantitative verification of the de Broglie relation

$$\lambda = h/mv$$

as regards equally the dependence of λ on *both the mass and the velocity*, and the absolute value of the wavelength.

Johnson's Experiments. The diffraction of atomic hydrogen at LiF has been studied by Johnson.[b] A beam of atomic hydrogen of circular or nearly circular cross section falls on the cleavage surface of the crystal, and the diffracted beams are detected at a MoO_3 target. Thus the results are qualitative only, but have nevertheless an elegance of their own, because the whole diffraction pattern can be observed at once on the target.

Plate III, Fig. 40 is a photograph of the diffraction pattern obtained from a beam of atomic hydrogen incident at 45° on a cleavage surface of LiF oriented as in Fig. 36 *b*. It should be compared with Fig. 35. At the centre is seen the dense spot due to the directly reflected beam; the parabola is the intersection of the α-cone with the target; the vertical straight line is the intersection with the target of the first β-cone, which for this orientation of the crystal degenerates to the xz plane. The positions of maximum intensity on the parabola, corresponding to $\lambda = \lambda_m$, are clearly distinguishable in the photograph.

[a] This point has been given some prominence, because the very fact of the occurrence of the discrepancies, far from being a flaw in the observations (cf. for example Kikuchi, *Physikal. Z.* 31, 777, 1930), is in beautiful agreement with the theory of the cross grating.

[b] Johnson, *Phys. Rev.* 35, 1299, 1930; *J. Franklin Inst.* 210, 135, 1930.

Plate III, Fig. 41 shows the diffraction pattern obtained when the beam is incident normally on the crystal. The beam passes through a hole in the detecting plate, falls perpendicularly on the crystal, and the directly reflected beam, which passes back through the hole in the target, is seen as a small black disc in the photograph. The two heavy lines crossing the central disc run at 45° to the cleavage edges of the crystal, and therefore represent the intersection with the target of the first α- and β-cones (which for normal incidence degenerate to the yz and xz planes respectively) arising from the grating formed by the rows of similar ions on the surface. At 45° to these lines may just be detected a second pair of much fainter lines, running parallel to the cleavage edges.[a] This is a particularly interesting result, because it shows that it is possible to obtain definite if weak cross-grating diffraction spectra from the grating formed by the rows of alternately dissimilar ions.[b]

REFLECTION OF CADMIUM AND ZINC FROM ROCKSALT

Ellett, Olson and Zahl[c] reported that beams of cadmium reflected from rocksalt were "monochromatic" to within ± 50 m./sec., the velocity (or in other words, the de Broglie wavelength) of the reflected beam depending on the angle of incidence. Such a selective effect is *not* the property of a two-dimensional grating; it is a characteristic of a *three-dimensional grating*, represented by the well-known Bragg formula

$$2a \sin \theta = n\lambda.$$

(See particularly P. P. Ewald, *Kristalle und Röntgenstrahlen*, Berlin, 1923, p. 56.)

[a] On the original photograph these lines, although faint, are distinguishable at a glance.

[b] Note added in proof: The patterns of Plate III, together with other interesting details, have been published by Johnson in the *Physical Review*, 37, 847, 1931.

[c] Ellett, Olson and Zahl, *Phys. Rev.* 34, 493, 1929.

Ellett and Olson[a] have made a series of determined attempts to fit their data into modified Bragg formulae. Recent experiments of Zahl[b] on cadmium and zinc have however failed to reproduce the earlier results. It may be mentioned that Johnson[c] found no trace of velocity selection in a beam of atomic hydrogen specularly reflected from rocksalt.

All that can be said with certainty about the results of Ellett, Olson and Zahl is that there is a seemingly sporadic appearance of what appears to be a partial velocity selection in beams of Cd and Zn reflected from NaCl. That a velocity selection could arise from the diffraction of the incident beam at a space grating is improbable; it would mean that an atom moving with thermal velocity could penetrate to a depth of several atomic layers into a solid, to return to the surface and escape, all without loss of energy. On the other hand, it is certain that for the very short de Broglie wavelengths of Cd and Zn (7·67 and 8·15 × 10⁻¹⁰ cm. respectively at the temperatures of the experiments) the $(0, \pm 1)$ orders *from a surface grating* would not be resolved from the $(0, 0)$ order; and the earlier experiments of Knauer and Stern[d] with unresolved diffraction patterns show indeed how confusing, and how difficult of interpretation are the results when such is the case.

[a] Ellett and Olson, *Science*, **68**, 89, 1928; Ellett, Olson and Zahl, *loc cit.*; Ellett, *Phys. Rev.* **35**, 293, 1930 (Abs.).

[b] Zahl, *Phys. Rev.* **36**, 893, 1930.

[c] Johnson, *J. Franklin Inst.* **207**, 693, 1929.

[d] Knauer and Stern, *Z. Physik*, **53**, 782, 1929 (U. z. M. 11).

Chapter 5

THE MAGNETIC DEVIATION OF
MOLECULAR RAYS

If a magnetic doublet, moment μ, is placed in a magnetic field H, there is a change of energy of amount

$$\Delta E = \mu_i H \qquad \ldots\ldots(5\cdot1),$$

where μ_i is the component of μ in the direction of H. If we take the centre of gravity of the doublet to be the origin of an arbitrary system of co-ordinates x, y, z, then the force acting on the doublet in the x direction is

$$F_x = \frac{\partial}{\partial x}(\mu_i H)$$

$$= \mu_i \cdot \frac{\partial H}{\partial x} + H \cdot \frac{\partial \mu_i}{\partial x} \qquad \ldots\ldots(5\cdot2).$$

If now the magnetic moment μ is associated with an angular momentum, and provided that the doublet cannot gain energy from sources, such as radiation or collision, other than the field, then quite generally

$$\mu_i = \text{constant},$$

and immediately from $(5\cdot2)$

$$F_x = \mu_i \frac{\partial H}{\partial x}.$$

Similarly

$$\left.\begin{aligned} F_y &= \mu_i \frac{\partial H}{\partial y}, \\ F_z &= \mu_i \frac{\partial H}{\partial z} \end{aligned}\right\} \qquad \ldots\ldots(5\cdot3).$$

It is clear from $(5\cdot3)$ that when the field is uniform $F_x = F_y = F_z = 0$. On the other hand, when the field is inhomogeneous, there exists a translatory force on the doublet, and a beam of molecules possessing a magnetic moment, when shot through such a field, will suffer a *deviation*, the extent and character of which enable us to draw important conclusions about the magnetic properties of the molecules constituting the beam. The interpretation will be complicated for the general case of diatomic and polyatomic molecules, except in exceptionally favourable cases, by the fact of temperature rotation; for this reason, and also because so far no adequate experimental data on molecules exist, we shall confine our attention here to the study of the magnetic deviation of *atomic* beams.

THE FIELD

The Stern-Gerlach Field. The choice of a suitable field is clearly of first importance for the practical realisation of a deflection method. The simplest arrangement is in principle that for which H and the gradient of the field have the same direction, say that of z.

This condition is fulfilled near the wedge-shaped pole-piece of an electromagnet, strictly in the plane of symmetry through the knife-edge and to a sufficient approximation for a short

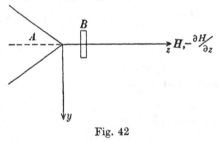

Fig. 42

distance on each side of that plane (cf. Fig. 42, where A is the pole-piece, and the xz plane is the plane of symmetry). With this arrangement, $F_x = F_y = 0$, and the atoms of a beam shot parallel to the edge of the pole-piece, that is in a direction perpendicular to the plane of the paper (cross-section at B, Fig. 42), suffer an acceleration at right angles to the beam of amount

$$f = \frac{F_z}{m} = \frac{\mu_i}{m}\frac{\partial H}{\partial z},$$

where m is the mass of an atom. The deflection s for given μ_i is then

$$s = \tfrac{1}{2}ft^2 = \tfrac{1}{4} \cdot \frac{F_z l^2}{E_v} \qquad \ldots\ldots(5\cdot4),$$

where t is the time of flight of an atom moving with a given velocity v, in a field of length l; and E_v is the energy of translation corresponding to the velocity v.

As a numerical example, we may take $l = 6$ cm., $F_z = \mu_i \dfrac{\partial H}{\partial z} \sim 10^{-20} \cdot 10^4$ (see below, p. 122); then, if v is taken to be the most probable velocity α, $E_\alpha = kT$; and we have, for $T = 1000°$ K.,

$$s = \frac{1}{4} \cdot \frac{10^{-16} \cdot 36}{1\cdot37 \times 10^{-16} \cdot 10^3} = 6\cdot6 \times 10^{-2}\,\text{mm}.$$

The arrangement of Fig. 42 presents rather serious disadvantages if an *absolute* determination of μ_i is aimed at. It will be clear from the numerical example just given that in order to obtain a deflection that is readily measurable, one must use a very narrow beam, say one $10–100\,\mu$ wide. Since it is essential to know the value of $\dfrac{\partial H}{\partial z}$ at the position of the beam, this involves a point-to-point plot of the field in the restricted space between the pole-pieces, at intervals of about $0\cdot1$ mm., both in the plane of symmetry, and for about 1 mm. on each side of it.

The evaluation of $\dfrac{\partial H}{\partial z}$ at any point in the field can be arrived at by making two separate series of measurements, of $H \cdot \dfrac{\partial H}{\partial z} = \text{grad } H^2$, and of H itself. The measurement of H at any point is very simply accomplished by determining the change of resistance suffered by a short bismuth wire about $0\cdot1$ mm. in diameter when placed at that point in the field; the wire having been previously calibrated in a homogeneous field (that is between flat pole-pieces) against a standard bismuth spiral. The most satisfactory method of measuring grad H^2, due to Stern and Riggert, is to observe the deflection in the z direction suffered by a quartz reed, to the end of which is attached a tiny test body of bismuth. For if κ is the volume susceptibility, v the volume of the test body, the z component of the force acting on it at any point of the field is

$$F_z = \kappa v . H \frac{\partial H}{\partial z} .$$

If the reed is previously calibrated with milligram weights, its deflection at any point in the field gives F_z directly in dynes. Moreover, the measurements can be freed from the admitted uncertainty of the susceptibility of bismuth by calibrating the values of $\frac{\partial H}{\partial z}$ in the plane of symmetry, obtained by the separate determination of grad H^2 and H, against the inhomogeneity obtained directly from measurements of ΔH_z and Δz in that plane.[a] Nevertheless, the measurements are extremely laborious, and the highest attainable accuracy is probably not more than 1 or 2 per cent.

The Rabi Field. A notable advance was made in 1929 by Rabi,[b] who showed that it was possible to obtain easily measurable deflections by sending the beam at an angle between *flat* pole-pieces (see Fig. 43). The deflection arises in the restricted region of inhomogeneity at the entrance to the field, where the lines of force run similarly to those for a parallel

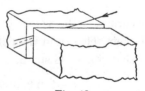

Fig. 43

plate condenser (see Fig. 66, Chapter 6). Since, however, the beam completely traverses the inhomogeneous zone it is only necessary to measure the *total* change in the field strength over the path of the beam. Now it is easily possible to

[a] Cf. Leu, *Z. Physik*, **41**, 551, 1927 (U. z. M. 4).
[b] Rabi, *Nature*, **123**, 163, 1929.

measure field strengths of, say, 10,000 gauss correct to 1 in 1000, so that as far as the field is concerned the Rabi method does indeed offer the possibility of making absolute and comparatively precise measurement of μ_i.

A full analysis of the Rabi method will be found at the end of this chapter; we give here a simple derivation which suffices if the accuracy aimed at is not greater than a few per cent.

Let a beam of atoms possessing a magnetic moment μ be sent at glancing angle θ into a field between flat pole-pieces,

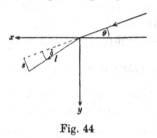

the boundary of which, supposed sharp, is defined by the x axis of a right-angled system of co-ordinates (Fig. 44). In traversing the infinitesimal region of inhomogeneity at the entrance to the field, the atoms will be accelerated normally to the boundary: we shall

Fig. 44

confine our attention for the present to those accelerated in the $+y$ direction. Since the acceleration is wholly normal to the boundary of the field

$$v_0 \cos \theta = v \cos (\theta + \delta),$$

where v_0 is the velocity of the atoms in field-free space and v their velocity in the field after acceleration at the entrance. Then, since δ is small,

$$1 - \tan \theta \tan \delta = \frac{v_0}{v} = E_0^{\frac{1}{2}} (E_0 + \mu_i H)^{-\frac{1}{2}} \quad ...(5\cdot5),$$

where E_0 is the kinetic energy corresponding to the velocity v_0.

Since $\frac{\mu_i H}{E_0}$ is small $\left(\sim \frac{10^{-20} \cdot 10^4}{10^{-13}} = 10^{-3} \right)$, (5·5) becomes

$$\delta = \frac{\mu_i H}{2E_0 \tan \theta}.$$

The deflection measured at a distance l from the entrance of the field in a direction perpendicular to the direction of the parent beam is then

$$s = l\delta = \frac{\mu_i}{2E_0} \cdot \frac{Hl}{\tan \theta} \qquad \ldots\ldots(5\cdot6),$$

where it is understood that the field extends to the position of the detector.

In practice, when the pole-pieces are a finite distance apart, the physical boundary of the field is not sharp, and H in (5·6) must be replaced by the average value $\bar{H}$ of the field strength over the path of the beam, giving

$$s = \frac{\mu_i}{2E_0} \cdot \frac{\bar{H}l}{\tan \theta} \qquad \ldots\ldots(5\cdot6').$$

It appears at first sight that the Rabi arrangement could be made exceedingly sensitive by choosing a very small glancing angle. This is, however, not the case; for in order to obtain the full deflection with the Rabi arrangement the beam must traverse completely the region of inhomogeneity at the entrance to the field. Now a field plot made at the field entrance showed that the extent Δ of this region is about three times the distance d between the pole-pieces—$2d$ above and d below the geometrical boundary of the field;[a] on the other hand, d cannot usefully be made smaller than about 2 mm. without unduly sacrificing intensity. In practice therefore $\Delta \sim 6$ mm., and geometrical considerations show at once that if the beam is to traverse this region completely, the length of path between image slit and detector becomes for small θ inordinately long, with a consequent loss in intensity far outbalancing any gain in deflection.

Thus the Rabi method is, unless quite unusual magnetic resources are available, actually rather less sensitive than the Stern-Gerlach method for a beam of given height; and so long as attention is restricted to *comparative* measurements against a standard substance, when a field plot in the latter case is no

[a] Rabi, *Z. Physik*, 54, 190, 1929 (U. z. M. 12).

longer necessary, the two methods are of the same intrinsic accuracy.

The Hamburg Set-up. The Stern-Gerlach arrangement is often the more convenient; but to get the best results from it the field must be rather carefully proportioned. Fig. 45 shows the typical form and dimensions of a Stern-Gerlach field designed to give the necessary inhomogeneity of 10^4 gauss/cm. The beam is shown in cross section just at the entrance of a deep narrow channel cut in the pole-piece opposite the wedge. This position has been chosen as the result of detailed field plots, which have shown that in the plane of the channel entrance the inhomogeneity over the height of the beam

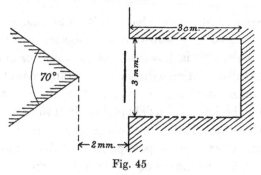

Fig. 45

(ca. 2 mm.) is constant to within a few per cent. in a well-proportioned field. This is clearly shown in Fig. 46, where the $\frac{\partial H}{\partial z}$: z curves for different values of y converge very closely immediately before the entrance of the channel.[a] The deflected traces arising from a parent beam sent through the field nearly in the plane of the channel entrance are therefore appreciably straight, instead of strongly bowed as results when the beam is sent near the wedge. Moreover, $\frac{\partial H}{\partial z}$ near the channel does not vary rapidly in the z direction, and a correction for a change of inhomogeneity along the path of a

---

[a] I am indebted to Dr J. B. Taylor for these particular data.

deflected component, which is essential in the neighbourhood of the wedge, is no longer necessary.[a]

In the sequel, the arrangement whereby the beam is sent through the field near the wedge will be referred to as the *Stern-Gerlach set-up*; that whereby the beam is sent near the

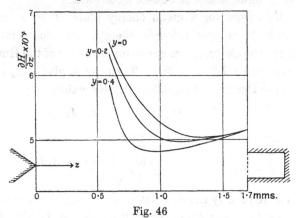

Fig. 46

channel, the *Hamburg set-up*. The features just mentioned are observable in a comparison of Plate IV, Fig. 47 *b*, which shows the deflection pattern obtained when a silver beam is sent near the wedge (Stern-Gerlach set-up); and Plate I, Fig. 6 *b*, showing that of a lithium beam sent near the channel (Hamburg set-up).

MAGNETIC DEFLECTION PATTERNS

We have next to consider quite generally what sort of deflection pattern is to be expected for a given type of atom. Now quantum theory allows certain discrete values only of μ_i in equation (5·1), which are defined by the relation

$$\mu_i = Mg \cdot \mu_B \qquad \ldots\ldots(5\cdot7),$$

[a] Cases may arise when it is desirable to send the beam as near the *wedge* as possible. Thus for example if it is intended to *detect* the existence of a rather small moment, it is best to send a beam defined by *long* slits very near the wedge: only the central portion of the beam will then be deflected, and the undeflected straight portions on either side of the centre serve as a very convenient basis of reference.

where M is the magnetic quantum number, g the Landé factor, and μ_B the quantum unit of magnetic moment, namely

$$\mu_B = eh/4\pi m_0 c = 0\cdot918 \times 10^{-20} \text{ erg gauss}^{-1},$$

in which e, m_0 are the electronic charge and mass, c is the velocity of light, and h is Planck's constant.

The Mg-values for a given energy state of an atom are predictable from spectroscopic theory, provided that the coupling of the electronic momentum vectors is of the Russell-Saunders type ((LS) coupling). Thus, for a given ^rL_J state, M can take the $2J + 1$ equally probable values

$$- J, \ - J + 1, \ ... \ J - 1, \ J,$$

and g is defined by the general relation

$$g = 1 + \frac{J(J + 1) + S(S + 1) - L(L + 1)}{2J(J + 1)}.$$

If, as is frequently the case with the heavier atoms, we have Stoner (or (jj)) coupling, then although the J-value for a given state, and hence the M-values, are identical with those obtaining with (LS) coupling, the g-value is different, and is no longer calculable from a general formula, but can be evaluated only in isolated favourable cases.

In Table VI are given the Mg-values, with the corresponding deflection patterns for a single velocity beam,[a] to be expected on the basis of Russell-Saunders coupling for a number of typical normal states, to which states alone (see p. 4) the molecular ray method is readily applicable. For Stoner coupling, where the g-values are in general not predictable, the deflection patterns are qualitatively the same as for Russell-Saunders coupling, but the magnitude of the separations is different.

The Stern-Gerlach Experiment. The deflection patterns in Table VI are in sharp contrast to those which the classical theory would have led one to expect. For on the classical

[a] The reason for this proviso will be apparent later.

Table VI

Normal state	g	Mg	Deflection pattern
$^1S_0$	$\frac{0}{0}$	0	
$^1P_1$	1	$-1\ \ 0\ +1$	
$^1D_2$	1	$-2\ -1\ \ 0\ +1\ +2$	
$^2S_{\frac{1}{2}}$	2	$-1\ \ +1$	
$^2P_{\frac{1}{2}}$	$\frac{2}{3}$	$-\frac{1}{3}\ +\frac{1}{3}$	
$^2P_{\frac{3}{2}}$	$\frac{4}{3}$	$-\frac{6}{3}\ -\frac{2}{3}\ +\frac{2}{3}\ +\frac{6}{3}$	
$^3S_1$	2	$-2\ \ 0\ +2$	
$^3P_0$	$\frac{0}{0}$	0	
$^3P_1$	$\frac{3}{2}$	$-\frac{3}{2}\ \ 0\ +\frac{3}{2}$	
$^3P_2$	$\frac{3}{2}$	$-\frac{6}{2}\ -\frac{3}{2}\ \ 0\ +\frac{3}{2}\ +\frac{6}{2}$	
$^4S_{\frac{3}{2}}$	2	$-3\ -1\ \ +1\ +3$	
$^5S_2$	2	$-4\ -2\ \ 0\ +2\ +4$	

theory μ_i could take any value continuously between 0 and μ, namely

$$\mu_i = \mu \cos (\mu H) = \mu . \cos \alpha.$$

Therefore in the deflection experiment the deflecting force in the direction s, namely $F = \mu \cos \alpha \frac{\partial H}{\partial s}$, would take all values from 0 to $\mu . \frac{\partial H}{\partial s}$ continuously. Now the intensity of the deflected beam for a given deflection s is dn/ds, where dn is the number of atoms suffering a deflection between s and $s + ds$. For a given velocity $s \propto F \propto \cos \alpha$; hence $ds \propto \sin \alpha\, d\alpha$. But $dn \propto \sin \alpha\, d\alpha$ also. Therefore dn/ds is constant. Thus a single velocity beam of *any* atomic type whatsoever would on the classical theory show the same type of deflection pattern—a broadened trace of uniform intensity. But a beam as ordinarily realised has a velocity distribution corresponding to the Maxwell distribution of the velocities of the atoms in the source; hence in this case the deflection pattern for all atoms alike would be a broadened trace, arising from the superposition of single velocity traces, with a *maximum* at the position of the undeflected beam; for there all velocities contribute to the intensity.

It will be shown (p. 129) that provided the deflection s_a corresponding to the most probable velocity a is greater than $3a$ ($2a$ is the width of the undeflected trace), a beam of atoms in a state possessing two Mg-values will split into two, even when the Maxwell distribution is taken into account. All the metals of Group I, for which the normal state is $^2S_{\frac{1}{2}}$, are included in this case. Thus a beam of silver atoms, formed in the ordinary way, will split into two when shot through a magnetic field of sufficiently great inhomogeneity. There is thus an intensity *minimum* at the position of the undeflected trace; whereas, as we have seen, the classical theory would predict under similar conditions a *maximum* at that point.

It was this clear-cut distinction between the predictions of the quantum theory on the one hand and of the classical theory on the other which first led Stern[a] to suggest the deflection experiment with silver as a crucial test between them. The experiment was carried out by Gerlach and Stern[b] in 1921–22. In the then primitive stage of the technique an oven with a circular aperture was used, with two collimating slits to define the beam. The intensity of the resulting beam was of course extremely feeble; and after exposures of eight hours, development by the wet method was still necessary to bring up the deflection pattern. The pattern obtained is seen in Plate IV, Fig. 47. Fig. 47 *a* shows the trace without field, Fig. 47 *b* the deflection pattern obtained with field. The parent beam has split into two discrete beams, which are, owing to the Maxwell distribution of the atoms in the source, broader than the undeflected beam. One has been deflected towards,

[a] Stern, *Z. Physik*, 7, 249, 1921.
[b] Gerlach and Stern, *Ann. Physik*, 74, 673, 1924.

Plate IV

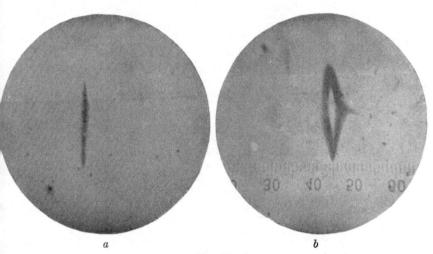

a   *b*

Fig. 47

The Stern-Gerlach Experiment

the other away, from the wedge-shaped pole-piece (cf. Fig. 42). The two beams are of equal average intensity. There is no trace of undeflected atoms.

This result, even in its purely qualitative aspect, was sufficient to decide unambiguously in favour of the quantum theory. Gerlach and Stern were however able to clinch the argument by making a rough measurement of the magnitude of the deflection. They found μ_i to be one Bohr magneton, in full agreement with spectroscopic theory.

It was however at first by no means clear how the silver atoms in the Stern-Gerlach experiment came to take up their definite "orientations" in the field.[a] The difficulties arose mainly out of a too concrete representation of the orientation process as a swinging into position of quasi-rigid magnets; a vector atom model is here unable to represent the facts of experiment. We have in fact to free ourselves entirely from pictorial representations, when it is no longer necessary to imagine that a sudden "change of orientation" takes place. Instead, we confine our attention to *directly observable* quantities, namely the energies $E_{\mathrm{mag}} = \mu_i H$ of the atoms in the states defined by the presence of the field, and assert merely the existence of a definite probability distribution of the atoms among the different energy states.

It is moreover impossible *in principle* to follow the change of state of an atom entering the field.[b] The time required for the atoms to enter the states defined by the field is of the order of the period τ of the Larmor precession, namely

$$\tau = 4\pi m_0 c/eH = 7 \times 10^{-7}.1/H.$$

The highest possible accuracy δE with which the energy can be determined *during this time* is given by the uncertainty relation, namely

$$\delta E\,\delta t \sim h.$$

That is
$$\delta E \sim h/\tau \sim 10^{-20}H.$$

[a] Cf. Einstein and Ehrenfest, *Z. Physik*, 11, 31, 1922.
[b] Cf. Heisenberg, *Z. Physik*, 43, 172, 1927.

But this is of the same order as the total field energy

$$E_{mag} = \mu_i . H \sim 10^{-20}H,$$

so that the atoms cannot be observed during a transition from one magnetic energy state to another, but only so long as they are in the energy states defined by the imposed field.

Mg-values. The pioneer experiments of Gerlach and Stern gave immediate promise of a general method for the evaluation of *Mg*-values which should be independent of spectroscopy. The realisation of this aim has, however, proved much more difficult in the details than was at first apparent, and the deflection method must still rely very largely on spectroscopic theory for the interpretation of its results. The difficulties arise out of the fact that each deflected beam is spread out into an energy spectrum ($s \propto 1/v^2$). Thus, as we have seen in Fig. 47, even in the simplest case of a symmetrical splitting into two components, the deflected traces have a width which cannot be neglected in estimating the amount of the deflection; while certain of the individual components of the more complex deflection patterns may not be resolved at all. Some knowledge of the intensity distribution in the deflected beams is therefore essential to an evaluation of the μ_i (or *Mg*) values from the observed deflection patterns; moreover, the deflection method must rely on spectroscopic theory to decide the number and relative abundance of the components actually present in an unresolved deflected beam before the deflection patterns can be interpreted without ambiguity. These difficulties would not of course arise if the parent beam were sensibly of uniform velocity (cf. Table VI); at the same time, the loss in intensity which must inevitably follow monochromatisation would raise new, technical difficulties of detection.

INTENSITY DISTRIBUTION IN THE
DEFLECTED BEAMS[a]

The broadening of the deflected traces makes it necessary to refer the deflection to some standard of velocity, and a convenient standard is the most probable velocity α of the molecules in the source. Then if, for given μ_i, s_a is the deflection corresponding to α, we have from (5·3) and (5·4)

$$s_a = \frac{\mu_i}{4kT} \cdot \frac{\partial H}{\partial z} \cdot l^2;$$

or, if the detector is placed a distance l_2 from the end of the field, itself of length l_1,

$$s_a = \frac{\mu_i}{4kT} \cdot \frac{\partial H}{\partial z} \cdot l_1^2 \left(1 + \frac{2l_2}{l_1}\right)$$

$$\dots\dots(5\cdot8) \text{ (Stern-Gerlach)}.$$

Alternatively, by (5·6'),

$$s_a = \frac{\mu_i}{2kT} \cdot \frac{\bar{H}l}{\tan \theta} \quad \dots\dots(5\cdot9) \text{ (Rabi)}.$$

When, as is sufficient for most purposes, a comparative measurement of μ_i is made against a standard substance, whose resolved magnetic moment is μ^*, then

$$s_a = s_a^* \cdot \frac{\mu_i}{\mu^*} \cdot \frac{T^*}{T_i} \quad \dots\dots(5\cdot10),$$

where T_i, T^* are the source temperatures for the unknown and standard substance respectively.

In order to evaluate μ_i, it is necessary to determine s_a from the characteristics of the deflected traces. Now $s \propto 1/v^2$; hence, changing the independent variable in (2·4) from v to $s = s_a \cdot a^2/v^2$, we get, for a given value of μ_i with the *a priori* probability w_i,

$$I_i(s)\, ds = w_i I_0 e^{-s_a/s} \cdot s_a^2/s^3 \cdot ds \quad \dots\dots(5\cdot11),$$

as the intensity for deflections between s and $s + ds$ in a deflected component arising from a parent beam whose width

[a] See Stern, Z. *Physik*, **41**, 563, 1927 (U. z. M. 5).

is negligible in comparison with that of the deflected trace. The intensity $\sum_i I_i$ at the point s in a complex deflected trace arising from a parent of constant intensity I_0 and width $2a$ is then (cf. Chapter 2, p. 72)

$$\sum_i I_i = \sum_i w_i I_0 \left[e^{-y} (y+1) \right]_{s_a/s-a}^{s_a/s+a} \quad \text{for } s > a$$

$$\qquad\qquad\qquad\qquad\qquad\qquad\qquad \ldots\ldots(5\cdot12\,a),$$

$$\sum_i I_i = \sum_i w_i I_0 \left[e^{-y} (y+1) \right]_\infty^{s_a/a-s} + \sum_i w_i I_0 \left[e^{-y} (y+1) \right]_\infty^{s_a/a+s}$$

$$\text{for } s < a \quad \ldots\ldots(5\cdot12\,b).$$

The Case of $w_i = \frac{1}{2}$. We shall exemplify in the first instance the method of evaluation of s_a from the characteristics of the deflected trace with reference to the simple case of symmetrical splitting into two beams ($w_i = \frac{1}{2}$); it being clearly understood that in so doing we are treating a particular case, the results of which are typical, but not necessarily general.

The intensity distribution for a single component is then by (5·12 a)

$$I = \tfrac{1}{2} I_0 \left[e^{-s_a/s+a} \left(\frac{s_a}{s+a} + 1 \right) - e^{-s_a/s-a} \left(\frac{s_a}{s-a} + 1 \right) \right]$$

$$\qquad\qquad\qquad\qquad\qquad\qquad\qquad \ldots\ldots(5\cdot13).$$

If $s \geqslant a$, this becomes, developing to first powers of a/s,

$$I = I_0 . a/s . e^{-s_a/s} . (s_a/s)^2 \quad \ldots\ldots(5\cdot14),$$

an approximate form which is very useful when it is desired to plot the intensity distribution graphically in making a preliminary survey of the results.

In Fig. 48, Curve I is drawn using (5·14) for the case $s_a = 10a$, on the same scale as the undeflected trace Curve O, which represents the form of the beam for the case of Fig. 10, p. 51, where the width of the image slit is twice that of the oven slit, and their distance apart is equal to the distance of the image slit from the detector; in calculating Curve I the intensity is taken to be constant over the undeflected trace, which is then represented by the dotted rectangle. Curve I is repeated in Curve II with the scale of ordinates ten times as large, in order that the general features of the deflected trace may be more clearly seen.

First it will be observed that the deflection s_a corresponding to the most probable velocity a lies far out from the position of maximum intensity s_m. From the figure $s_m \cong \frac{1}{3} s_a$; this is

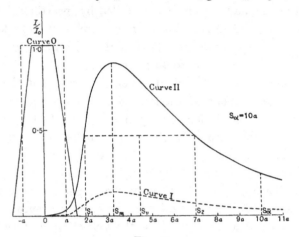

Fig. 48

at once controllable from (5·13), for from the condition $dI/ds = 0$ we obtain

$$s_a = \frac{s_m{}^2 - a^2}{2a} \cdot 3 \log \frac{s_m + a}{s_m - a} \qquad \ldots\ldots(5\cdot15)$$

$$\cong 3 s_m \text{ for } a \ll s_m.$$

Thus a *minimum* at $s = 0$ is first obtained when

$$s_a > 3a \ (s_m > a),$$

a result which follows directly from (5·12 b).

Secondly, we note that the deflection s_v corresponding to the *centre* of the deflected trace bears no simple relation to s_a; in fact, the ratio s_v/s_a varies with s_a and with a. Specific attention was first drawn to this by Semenoff.[a] The point is of

[a] Semenoff, *Z. Physik*, **30**, 151, 1924.

historical interest, because Gerlach and Stern, in evaluating their results with silver, based themselves on the velocity measurements of Stern (see Chapter 2, p. 63), and took s_v to correspond to $v = \sqrt{3 \cdot 5kT/m}$. The error involved in this assumption was however, with the comparatively wide beams and small deflections obtaining in their experiment, of the same order as the experimental accuracy (ca. 10 per cent.) aimed at, as may readily be controlled by insertion of their data in (5·8) and (5·13). For narrow beams and large deflections, such as are now practicable, the corresponding error would be very large (20 to 40 per cent.).

Evaluation of s_a. It would appear at first sight that intensity measurements in the deflected trace are necessary to evaluate s_a. This is fortunately not necessarily the case— fortunately, because only the non-quantitative target detector is available at present for the vast majority of atoms. The following method has proved itself reliable: The limits of visibility s_1, s_2 of the trace on the target are measured, either directly with ocular micrometer, or more accurately by photometry of a photograph of the deflection pattern.[a] If I_1, I_2 are the intensities at s_1, s_2, then clearly $I_1 = I_2$, and from (5·13)

$$\left[e^{-s_a/s_1 + a} \left(\frac{s_a}{s_1 + a} + 1 \right) - e^{-s_a/s_1 - a} \left(\frac{s_a}{s_1 - a} + 1 \right) \right]$$

$$= \left[e^{-s_a/s_2 + a} \left(\frac{s_a}{s_2 + a} + 1 \right) - e^{-s_a/s_2 - a} \left(\frac{s_a}{s_2 - a} + 1 \right) \right]$$

$$\ldots\ldots(5\cdot16),$$

from which s_a is obtained by trial and error. In case $a/s_1 \ll 1$

---

[a] This statement is not in conflict with the conclusion arrived at in Chapter 1, p. 32, that direct photometry of a deposit as a means of measuring intensities is unreliable; for here we are concerned merely with the determination of positions of *equal* intensity, a measurement which does not depend on a knowledge of the characteristic curve.

(in practice $< \frac{1}{2}$), the expressions in square brackets can be developed in terms of a/s, and one obtains

$$s_a = \frac{3s_1 s_2}{s_2 - s_1} \left(\log \frac{s_2}{s_1} + \epsilon \right) \Bigg\} \quad ...(5 \cdot 17),$$
$$\epsilon = \tfrac{1}{3} . a^2/s_1{}^2 . (2 - \tfrac{4}{3} . s_a/s_1 + \tfrac{1}{6} . s_a{}^2/s_1{}^2)$$

where terms in a/s_1 higher than the second, those in a/s_2 higher than the first power have been neglected. It will frequently be found in practice that the correction factor ϵ amounts to less than one per cent. of s_a, and can be neglected.

Table VII

Potassium

Stern-Gerlach Field, Hamburg Set-up. Width of Channel 4 mm. $l_1 = 6$ cm. $\frac{\partial H}{\partial z_a} = 4 \cdot 15 \times 10^4$ gauss/cm.; $\frac{\partial H}{\partial z_b} = 4 \cdot 10 \times 10^4$ gauss/cm. (cf. Note on entries in column 7). $T = 687°$ K. Target material: Glass, chemically silvered.

1	2	3	4	5	6	7	8	9	10
No.	l_2 in cm.	Time of appearance of Trace	Measured after	a in μ	s_1 in μ	s_2 in μ	s_a from (5·37)	s_a from (5·29)	% age difference
		hr. min.	hr. min.						
1	2·7	— 20	2 —	30	142	382	670	653	+ 2·6
2 a	2·7	— 19	2 30	30	140	390	670	659	+ 1·7
2 b						370	657	648	+ 1·4
3 a	5·7	1 30	3 —	32	195	645	1000	1000	0
3 b						620	986	988	− 0·2
4 a	5·7	1 10	2 30	32	195	645	1000	1000	0
4 b						620	986	988	− 0·2
5 a	5·7	1 40	2 30	32	190	635	981	1000	− 1·9
5 b						590	951	988	− 3·7

Table VII exemplifies the use of this method of evaluation in practice. The data are taken from a paper by Leu.[a] In column 7, note that two values of s_2 are entered for all experiments except the first, one referring to the component of the deflection pattern which has been deflected away from,

[a] Leu, *Z. Physik*, **41**, 551, 1927 (U. z. M. 4).

the other to that deflected towards the wedge-shaped pole-piece; it is clear from the table that the former is the more strongly deflected. This asymmetry in the deflection pattern, which may be seen very clearly in Fig. 6 *b*, frequently arises with the Hamburg set-up; for although the inhomogeneity near the channel is tolerably constant over the width *of an individual component* (see p. 120 above), it is not constant over the entire deflection pattern if the deflections are large enough, but is greater for the components deflected towards the channel (cf. Fig. 46). s_a in column 8 is calculated from (5·17) with the values of s_1, s_2 given in columns 6 and 7. s_a in column 9 is calculated from (5·8) on the assumption that μ_i is one Bohr magneton.

The Case of $w_i \neq \frac{1}{2}$. When $w_i \neq \frac{1}{2}$ we no longer have the parent beam split into two symmetrically deflected beams by the field. If the atoms emerging from the source all moved with the same speed, this would offer no complication; deflection patterns like those of Table VI would be obtained, which would be immediately interpretable. As we have seen, the facts are otherwise: the atoms in the deflected beams are spread out into an energy spectrum, and the deflected traces are consequently so wide that it is not in general possible to obtain the full number of deflected components.

We shall for the sake of definiteness discuss a particular case, namely when the J-value of the normal state is $\frac{3}{2}$ (e.g., $^4S_{\frac{3}{2}}$ in Table VI). Then with a monochromatic beam one would obtain a deflection pattern of *four* equally intense traces, corresponding to $\mu_i = \pm \frac{1}{2}g$, $\mu_2 = \pm \frac{3}{2}g$; thus w_i is here $\frac{1}{4}$. When the velocity distribution of the atoms in the beam is taken into account, the intensity distribution in the deflected traces has the general character of the broken curves in Fig. 49, which represents the right-hand half only of the complete deflection pattern. The summation curve of the two components is the full curve in the figure; it is evident that it has but a single maximum. Thus actually the full deflection pattern consists of only *two* traces, situated symmetrically right and left of the position of the undeflected trace.

It is however possible to calculate the μ_i-values from the characteristics of the summation curve, *provided that the J-value is known*. Thus in the present example, where the value $J = \frac{3}{2}$ has been assumed, we know the summation curve, which alone is amenable to observation, to be the resultant

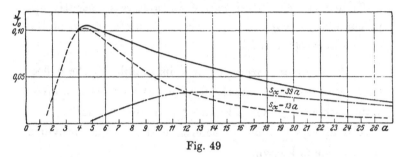

Fig. 49

of two curves corresponding to μ_1, $3\mu_1$. Thus the equation to the summation curve (for the case $s_a \gg a$) is, from (5·14),

$$I_{\text{summation}} = \tfrac{1}{2}I_0 a\,(e^{-s_a/s}.s_a{}^2/s^3 + e^{-3s_a/s}.9s_a{}^2/s^3)$$
$$= \tfrac{1}{2}I_0 a.s_a{}^2/s^3\,(e^{-s_a/s} + 9e^{-3s_a/s}) \quad \dots(5\cdot18),$$

in which s_a has the meaning $s_a\,(\mu_1)$.

Assuming as before that the limits of visibility s_1, s_2 of the trace correspond to equal intensities I_1, I_2 we have

$$s_a{}^2/s_1{}^3\,(e^{-s_a/s_1} + 9e^{-3s_a/s_1}) = s_a{}^2/s_2{}^3\,(e^{-s_a/s_2} + 9e^{-3s_a/s_2})$$
$$\dots\dots(5\cdot19),$$

from which $s_a\,(\mu_1)$ can be determined.

Scope of the Deflection Method. It should be pointed out that the assumption of a symmetrical splitting into two beams, as in the Stern-Gerlach experiment with silver, implies a knowledge of the J-value (namely, $\frac{1}{2}$) in exactly the same degree as in the case just discussed. Thus *perfectly generally* we can say that it is impossible for the deflection method to determine μ_i-values independently of a knowledge of spectroscopic theory, *unless* a monochromator (q.v.) is used.

It is of course possible to know the J-value of a state from spectroscopic evidence without a simultaneous knowledge of the g-value, if the coupling of the electronic momentum vectors in the atom is not of the Russell-Saunders type (see above, p. 122). Since $\mu_i = Mg$, the deflection method *is* capable of measuring g-values of states for which the J-values are known, independently of spectroscopy.

These facts are unfortunately not always sufficiently appreciated; we therefore summarise the useful range of the deflection method as follows:

Without a monochromator, it is impossible for the deflection method to determine the μ_i-values of an atom in a given state without a previous knowledge of the J-value.

It is always possible to determine the g-value, provided the J-value is known.

If besides the normal state, *metastable states* are present in the beam (e.g., atomic oxygen) a knowledge merely of the J-values of all states occurring is not sufficient. In addition, the values of their statistical weights, as calculated from the energy differences of the states and the temperature of the source, are necessary before μ_i-values can be determined from the characteristics of the summation traces.

THE μ_i-VALUES OF THE ATOMS

Nineteen atomic species had been investigated by the deflection method up to the end of 1930. The results, which are on the average correct to some 3 to 5 per cent., confirm in the main the Mg-values for the normal states found or predicted spectroscopically; their scope can best be appreciated from a systematic discussion, with the Periodic Classification as guide. Table VIII gives a list of the normal states which have been determined by spectroscopic observations or predicted on the assumption of Russell-Saunders coupling.

Element	Term
—	
Ra	$^1S_0$
Ac	$^2D_{\frac{3}{2},\frac{5}{2}}$
Th	?
Pa	?
U	?

Element	Term
Cs	$^2S_{\frac{1}{2}}$
Ba	$^1S_0$
La	$^2D_{\frac{3}{2},\frac{5}{2}}$
Ce	^3H
Pr	^4K
Nd	^5L
Il	^6L
Sm	^7K
Eu	^8H
Gd	^9D
Tb	^8H
Ds	^7K
Ho	^6L
Er	^5L
Tm	^4K
Yb	^3H
Lu	$^2D_{\frac{3}{2},\frac{5}{2}}$
Hf	$^3F_{2,3,4}$
Ta	$^4F_{\frac{3}{2},\frac{5}{2},\frac{7}{2},\frac{9}{2}}$
W	$^5D_{0,1,2,3,4}$
Re	?
Os	?
Ir	$^4F_{\frac{9}{2},\frac{7}{2},\frac{5}{2},\frac{3}{2}}$
Pt	$^3F_{4,3,2}$
Au	$^2S_{\frac{1}{2}}$
Hg	$^1S_0$
Tl	$^2P_{\frac{1}{2},\frac{3}{2}}$
Pb	$^3P_{0,1,2}$
Bi	$^4S_{\frac{3}{2}}$
Po	$^3P_{2,1,0}$
—	
Rn	$^1S_0$

Element	Term
Rb	$^2S_{\frac{1}{2}}$
Sr	$^1S_0$
Y	$^2D_{\frac{3}{2},\frac{5}{2}}$
Zr	$^3F_{2,3,4}$
Nb	$^6D_{\frac{1}{2},\frac{3}{2},\frac{5}{2},\frac{7}{2},\frac{9}{2}}$
Mo	$^7S_3$
Ma	—
Ru	$^5F_{5,4,3,2,1}$
Rh	$^4F_{\frac{9}{2},\frac{7}{2},\frac{5}{2},\frac{3}{2}}$
Pd	$^1S_0$
Ag	$^2S_{\frac{1}{2}}$
Cd	$^1S_0$
In	$^2P_{\frac{1}{2},\frac{3}{2}}$
Sn	$^3P_{0,1,2}$
Sb	$^4S_{\frac{3}{2}}$
Te	$^3P_{2,1,0}$
I	$^2P_{\frac{3}{2},\frac{1}{2}}$
X	$^1S_0$

Element	Term
K	$^2S_{\frac{1}{2}}$
Ca	$^1S_0$
Sc	$^2D_{\frac{3}{2},\frac{5}{2}}$
Ti	$^3F_{2,3,4}$
V	$^4F_{\frac{3}{2},\frac{5}{2},\frac{7}{2},\frac{9}{2}}$
Cr	$^7S_3$?
Mn	$^6S_{\frac{5}{2}}$
Fe	$^5D_{4,3,2,1,0}$
Co	$^4F_{\frac{9}{2},\frac{7}{2},\frac{5}{2},\frac{3}{2}}$
Ni	$^3F_{4,3,2}$
Cu	$^2S_{\frac{1}{2}}$
Zn	$^1S_0$
Ga	$^2P_{\frac{1}{2},\frac{3}{2}}$
Ge	$^3P_{0,1,2}$
As	$^4S_{\frac{3}{2}}$
Se	$^3P_{2,1,0}$
Br	$^2P_{\frac{3}{2},\frac{1}{2}}$
Kr	$^1S_0$

Element	Term
Na	$^2S_{\frac{1}{2}}$
Mg	$^1S_0$
Al	$^2P_{\frac{1}{2},\frac{3}{2}}$
Si	$^3P_{0,1,2}$
P	$^4S_{\frac{3}{2}}$
S	$^3P_{2,1,0}$
Cl	$^2P_{\frac{3}{2},\frac{1}{2}}$
A	$^1S_0$

Element	Term
Li	$^2S_{\frac{1}{2}}$
Be	$^1S_0$
B	$^2P_{\frac{1}{2},\frac{3}{2}}$
C	$^3P_{0,1,2}$
N	$^4S_{\frac{3}{2}}$
O	$^3P_{2,1,0}$
F	$^2P_{\frac{3}{2},\frac{1}{2}}$
Ne	$^1S_0$

Element	Term
H	$^2S_{\frac{1}{2}}$
He	$^1S_0$

Hydrogen

Normal State, $^2S_{\frac{1}{2}}$; $Mg = \pm 1$

Atomic hydrogen has been investigated independently by Phipps and Taylor,[a] and by Wrede.[b]

The source of atomic hydrogen was in both cases a Wood's tube, from which the atoms emerged through a glass slit into a collimator chamber kept evacuated by fast pumps. A source slit of glass is necessary, because metal slits are strongly attacked by atomic hydrogen. Glass slits of ca. 0·05 mm. width can be made quite successfully by sealing a narrow strip of copper or steel foil into a glass pinch, and subsequently dissolving out the metal with acid. The target was MoO_3.

Measurement of the double trace obtained gave in both cases $\mu_i = \pm 1\mu_B$, in agreement with spectroscopy.

Wrede's measurements are by far the more accurate; he used the Hamburg set-up, and obtained therefore straight traces to which equations (5·8) and (5·17) were applicable. Estimated error 4 to 5 per cent.

Phipps and Taylor followed very closely the original Stern-Gerlach set-up. The beam was sent near the wedge; the deflection s_v of the middle of the deflected trace was taken, following Gerlach and Stern, to correspond to $v = \sqrt{3·5kT/m}$ (see p. 130 above); and finally, the field was not plotted, but Gerlach and Stern's values of $\frac{\partial H}{\partial z}$ were taken over on the ground that the pole-piece dimensions were similar. The errors so introduced could easily have amounted to as much as 20 per cent.

[a] Phipps and Taylor, *Phys. Rev.* **29**, 309, 1927.
[b] Wrede, *Z. Physik*, **41**, 569, 1927 (U. z. M. 6).

Group I a: The Alkali Metals
Normal State, $^2S_{\frac{1}{2}}$; $Mg = \pm 1$

Lithium. Investigated by Taylor,[a] using the Hamburg set-up. The deflection pattern obtained is reproduced in Plate I, Fig. 6. The measurements are comparative, against potassium as standard. $\mu_i = \pm 1\mu_B$, to 2 or 3 per cent.

This result is of particular interest from the spectroscopic side. Schüler[b] had observed hyperfine structure in the first line $^3S_1 - {}^3P_{012}$ of the triplet series of Li$^+$; the triplet separation was of the order 5 cm.$^{-1}$, the hyperfine structure separation of the order 0·5 cm.$^{-1}$—a quite exceptionally large hyperfine structure separation. This led Heisenberg[c] to suggest that the lithium nucleus possessed a magnetic moment of the order of a Bohr magneton (cf. pp. 147 ff. below). A nuclear magnetic moment of this magnitude, vectorially compounded with the electron spin moment of the neutral normal lithium atom, should have materially affected the μ_i value of the atom. Taylor's result however is in good agreement with the μ_i value predicted on the assumption of zero or negligibly small nuclear magnetic moment.

Goudsmit and Young[d] have recently shown that Taylor's result is in full agreement with the spectroscopic data, rightly interpreted. A direct comparison of the multiplet separation and hyperfine structure separation is in fact in this case unjustifiable, in that the former arises from the spin moment of the outer $2p$ electron, whereas the latter in the states $1s\,2s\,^3S$, $1s\,2p\,^3P$ arises from the deeply penetrating inner $1s$ electron. Calculation on this basis shows that the hyperfine structure separation arising from the interaction of a nuclear moment of the order $1/1840\mu_B$ on the $1s$ electron is of the observed order of magnitude.

[a] Taylor, *Z. Physik*, **52**, 846, 1929 (U. z. M. 9).
[b] Schüler, *Ann. Physik*, **76**, 292, 1925; *Z. Physik*, **42**, 487, 1927.
[c] Heisenberg, *Z. Physik*, **39**, 516, 1926.
[d] Goudsmit and Young, *Nature*, **125**, 461, 1930.

Sodium, Potassium. Investigated by Taylor,[a] with the Stern-Gerlach set-up, and by Leu[b] using the Hamburg set-up. Taylor's results are approximate only; he made the same assumptions about s_v and $\dfrac{\partial H}{\partial z}$ as did Phipps and Taylor with hydrogen (see above, p. 136).

Leu found in both cases $\mu_i = \pm\ 1\mu_B$; estimated error, $\pm\ 2$ per cent. for potassium, 3 to 4 per cent. for sodium.

Plate V, Fig. 50 shows a potassium deflection pattern obtained by Rabi,[c] in the course of orientating experiments on the practicability of his deflection method.

Sodium and potassium are likely to prove important for the deflection method; for an absolute determination of μ_i for these elements with an accuracy of at any rate 1 in 500 seems very hopeful (see below, pp. 150 ff.).

Group I b: Copper, Silver, Gold
Normal State, $^2S_{\frac{1}{2}}$; $Mg = \pm\ 1$

The resolved magnetic moment of silver was determined in the original work of Gerlach and Stern[d] as $\pm\ 1\mu_B$, correct to 10 per cent.

Comparative measurements on copper and gold, against silver as standard, were made by Gerlach.[e] He found in each case $\mu_i = \pm\ 1\mu_B$.

Group II b: Zinc, Cadmium, Mercury
Normal State, $^1S_0$; $Mg = 0$

Investigated by Leu;[f] in each case, the trace with field was of exactly the same width as that without field. The atoms of zinc, cadmium and mercury have zero magnetic moment.

[a] Taylor, *Phys. Rev.* **28**, 576, 1926.

[b] Leu, *Z. Physik*, **41**, 551, 1927 (U. z. M. 4).

[c] Rabi, *Z. Physik*, **54**, 190, 1929 (U. z. M. 12).

[d] Gerlach and Stern, *Ann. Physik*, **74**, 673, 1924.

[e] Gerlach, *Ann. Physik*, **76**, 163, 1925; see also Gerlach and Cilliers, *Z. Physik*, **26**, 106, 1924.   [f] Leu, *loc. cit.*

Group III. Thallium
Normal State, $^2P_{\frac{1}{2}, \frac{3}{2}}$

Thallium was investigated by Leu,[a] using the Hamburg set-up. $\mu_i = \pm \frac{1}{3}\mu_B$, correct to ± 4 per cent.

The normal state of the thallium atom is a regular doublet $^2P_{\frac{1}{2}, \frac{3}{2}}$ with the following relative term values (Paschen-Götze, *Seriengesetze der Linienspektren*, Berlin, 1922).

	$^2P_{\frac{1}{2}}$	$^2P_{\frac{3}{2}}$
$\Delta\nu$	0	7792·45 cm.$^{-1}$

The sub-level $^2P_{\frac{1}{2}}$ is stable; $^2P_{\frac{3}{2}}$ is metastable. Hence both levels are present in the beam with the relative probabilities

$$W = (2J + 1)\, e^{-hc.\Delta\nu/kT} = (2J + 1)\, e^{-1\cdot42\Delta\nu/T}$$
$$\dots\dots(5\cdot20).$$

At $T = 980°$ K., as in Leu's experiments, we obtain the following values:

	$^2P_{\frac{1}{2}}$	$^2P_{\frac{3}{2}}$
W	2	7×10^{-5}

Thus the metastable state is present to a negligible extent in the beam, and the Mg-value measured should refer to the $^2P_{\frac{1}{2}}$ level alone; namely $Mg = \pm \frac{1}{3}$, the value found by Leu.

Group IV. Tin, Lead
Normal State, $^3P_0$; $Mg = 0$

Investigated by Gerlach;[b] both have zero magnetic moment.

Group V. Bismuth
Normal State, $^4S_{\frac{3}{2}}$

The investigation of bismuth, with the Hamburg set-up, by Leu and the author,[c] brought out some points of great

---

[a] Leu, *loc. cit.*; see also Gerlach, *loc. cit.*
[b] Gerlach, *loc. cit.*
[c] *Z. Physik,* **49**, 498, 1928 (U. z. M. 8).

interest methodically. With $J = \frac{3}{2}$ we have *four* Mg-values, namely $Mg = \pm \frac{1}{2}g, \pm \frac{3}{2}g$. In agreement with the theory of p. 132, a beam arising from a vapour sufficiently strongly superheated to eliminate Bi_2 molecules (cf. Chapter 7, pp. 184 ff.) gave a symmetrical deflection pattern, not of four, but of *two* components[a] (see Plate VII, Fig. 72 *b*). Evaluation of the deflected traces using (5·19) gave $\mu_1 = 0\cdot72\mu_B \pm 3$ per cent. The corresponding Mg-value is $\frac{1}{2}g$, whence $g = 1\cdot45$.

Now the value of g predicted on the assumption of Russell-Saunders coupling is 2. It is however not surprising that a heavy atom like bismuth should depart from (LS) coupling. Fortunately, the g-value for extreme (jj) coupling is for bismuth predictable, as shown in the following scheme:

Bi : 3 valence electrons, $l_1 = l_2 = l_3 = 1$.

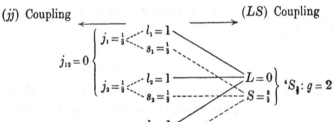

[a] Gerlach (*loc. cit.*) had previously investigated bismuth, with the Stern-Gerlach set-up, and obtained an unsymmetrical pattern consisting of a strong undeflected trace with a spur towards the wedge-shaped pole-piece. The asymmetry is probably not a real effect, but arises from the fact that Gerlach sent the beam near the wedge: the atoms attracted towards the pole-piece, particularly those with the larger component in the field direction, follow the strongly convergent lines of force near the wedge, and the effective intensity of the attracted beam is therefore increased; on the other hand, the atoms repelled from the wedge are dispersed. It is thus quite possible for the intensity of the attracted beam to be above, that of the repelled beam to fall below the transition intensity necessary for the formation of a permanent deposit. The strong undeflected trace is undoubtedly to be ascribed to the considerable proportion of molecules present at the temperature (ca. 700° K.) of the experiment (cf. Chapter 7, p. 184).

Plate 4.

Plate V

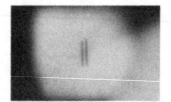

Fig. 50

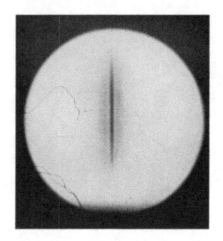

Fig. 51

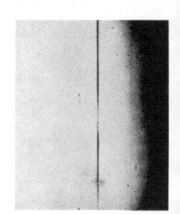

Fig. 54

Magnetic Deflection Patterns

Evidently, the g-value found for bismuth approaches quite closely the value for extreme (jj) coupling. It should be mentioned that the g-value for the normal state of bismuth has not so far been determined spectroscopically.

From the methodic standpoint, bismuth is the most typical example at once of the limitations and of the capabilities of the deflection method. Thus if the J-value had not been known spectroscopically, then on purely experimental grounds there would have been no reason for supposing the traces of Plate VII, Fig. 72 b other than simple. On that assumption a single μ_i-value of $0\cdot85\mu_B$ would have been obtained. On the other hand, with the J-value known, a hitherto undetermined g-value was measured. Thus we have exemplified in the case of bismuth an almost complete index of the deflection method as set out on p. 134.

Group VI. Oxygen

Normal State, $^3P_{2,1,0}$

Investigated by Kurt and Phipps,[a] using the Hamburg set-up, with an electrodeless discharge tube as the source of atomic oxygen, and a litharge target as detector. The deflection pattern obtained is reproduced in Plate V, Fig. 51. It consists of an undeflected trace and two deflected traces; the slight asymmetry arises from the difference in inhomogeneity at the position in the field of the attracted and repelled components (cf. Table VII, column 7).

Comparative measurements were made against atomic hydrogen ($\mu^* = 1\mu_B$), as standard. The traces were photographed, and photometered to determine s_1 and s_2.[b] The μ_i-values were then determined from (5·10), with μ^* set equal to $1\mu_B$, to be 0 and $1\cdot67\mu_B$.

---

[a] Kurt and Phipps, *Phys. Rev.* **34**, 1357, 1929.
[b] See footnote, p. 130.

The normal state of the O atom is an inverted triplet $^3P_{210}$ with the following relative term values:[a]

	$^3P_2$	$^3P_1$	$^3P_0$	
$\Delta\nu$	0	159	226	cm.$^{-1}$

The sub-level $^3P_2$ is stable, the sub-levels $^3P_{10}$ are metastable; hence all the levels are present in the beam, with the relative probabilities W given by (5·20), with $T = 613°$ K., namely

	$^3P_2$	$^3P_1$	$^3P_0$
W	5	2·07	0·59

The corresponding Mg-values, namely ± 3, $\pm \frac{3}{2}$, 0; $\pm \frac{3}{2}$, 0; 0; have therefore the relative probabilities $W(Mg)$ shown in the following table:

Mg	3	$\frac{3}{2}$	0
$W(Mg)$	0·438	0·741	1

Thus the deflected beams in Fig. 51 arise each from the superposition of two deflected beams, corresponding to $\mu_1 = \frac{3}{2}\mu_B$, $\mu_2 = 3\mu_B$ with the relative probabilities 0·741, 0·438 respectively. Fig. 52, taken from Kurt and Phipps' paper, shows the corresponding intensity curves, B and C in the figure, drawn with reference to the observed hydrogen curve A as standard. The summation curve D, just as with bismuth, has but a single maximum. The average value of μ_i calculated from the summation curve is $1·71\mu_B$, in excellent agreement with the value $1·67\mu_B$ derived from the observed traces.

Group VIII. Iron, Cobalt, Nickel

Iron. The normal state is $^5D_4$, with $Mg = \pm \frac{3}{2}$ (0, 1, 2, 3, 4). Investigated by Gerlach,[b] who found only a single undeflected trace with field. An undeflected trace would from the predicted Mg-values be expected; but deflected traces would also be expected. It is quite conceivable that possible de-

[a] Hopfield, *Astrophys. J.* **59**, 114, 1924.
[b] Gerlach, *loc. cit.*

flected traces could have escaped observation under the conditions of Gerlach's experiment. His intensities were very weak, and the target was removed from the apparatus for wet development of the invisible pattern. The development was never wholly successful, owing to the rapid oxidation of the tenuous iron deposit. It would then be quite possible for the more intense undeflected trace to escape, the less intense deflected traces to suffer destruction.

The experiment must therefore be regarded as inconclusive.

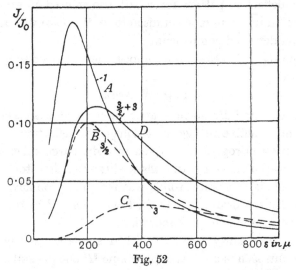

Fig. 52

Cobalt, Nickel. Cobalt and nickel have been investigated by Gerlach and his pupils.[a] In these experiments the beam was sent through the field approximately midway between wedge and channel.[b] The deflection patterns obtained display some hitherto unexplained features, inasmuch as there is an apparent resolution of the individual μ_i components. This is contrary alike to the analysis of p. 132, and to general experimental experience (examples, bismuth and atomic oxygen).

The normal state of nickel is $^3F_{432}$. At the temperature of

[a] Gerlach, *loc. cit.* (Ni); *J. de Physique*, VI, **10**, 273, 1929 (Co).
[b] Gerlach, *Ann. Physik*, **76**, 187, 1925.

Gerlach's experiments (ca. $1700°$ K.), $^3F_4$ is predominant, with the Mg-values $\pm \frac{5}{4}$ (4, 3, 2, 1, 0). One would expect, therefore, just as with atomic oxygen, a deflection pattern of three components—two composite traces symmetrically deflected right and left of a central undeflected trace. Gerlach obtained, however, in addition to a central undeflected trace and two approximately symmetrically deflected traces with $\mu_i \sim 1 \cdot 1 \mu_B$, indications of at least two further more strongly deflected discrete components.

The deflection pattern for cobalt also showed an apparent resolution into a number of discrete, in this case without exception deflected components.

The origin of these curious appearances is at present quite obscure.

Active Nitrogen

Jackson and Broadway[a] attempted to analyse the constituents of active nitrogen by the magnetic deflection method. The active nitrogen was formed in a discharge tube, and a beam was defined by two glass slits; the detector was a target of silver nitrate, which was reduced, probably to the oxide, by the beam. Atomic hydrogen, formed in the same apparatus, was used as standard.

It is now fairly well established spectroscopically that active nitrogen is a mixture of atomic ^2P and ^2D states, and the molecular $^3\Sigma$ state, all of which are metastable.[b] The corresponding Mg-values are

$^2P_{\frac{1}{2}}$	$^2P_{\frac{3}{2}}$	$^2D_{\frac{3}{2}}$	$^2D_{\frac{5}{2}}$	$^3\Sigma$
Mg $\pm \frac{1}{3}$	$\pm \frac{2}{3}, \pm 2$	$\pm \frac{2}{5}, \pm \frac{6}{5}$	$\pm \frac{3}{5}, \pm \frac{9}{5}, \pm 3$	$0, \pm 2$

One is dealing therefore with a very intricate case. There is the further complication that the effective temperature of the discharge was not known, and hence the statistical weights

[a] Jackson, *Nature*, 125, 131, 1930; Jackson and Broadway, *Proc. Roy. Soc.* 127, 678, 1930.

[b] See Kneser, *Ergebnisse der exakt. Naturwiss.* VIII, 1929, p. 229, for a summary of the active nitrogen problem.

of the several Mg-values are not calculable. Thus the experiments did not, *and could not*, yield definite confirmation of the spectroscopic data (cf. p. 134).

The deflection patterns, on twelve to twenty-four hours' exposure, were not resolved into two components, but a trace some 60 per cent. broader than without field was obtained. This poor resolution was forced upon the experimenters by the fact that unless the image slit was given a width incompatible with a definite splitting, the intensity was intractably small. Jackson and Broadway attempted to estimate the Mg-value for the broadened trace, interpreted as the unresolved trace of two simple components, by comparing its width with the total width of the hydrogen deflection pattern, the exposure in each case being prolonged until no further widening of the traces could be detected. This procedure is in principle indefensible, because it assumes what is not known to be the case: namely, that the critical beam intensity, below which a visible trace on a chemical target is not obtained even on infinitely prolonged exposure, is necessarily the same for all beam species. It is in fact absolutely essential for the evaluation (as distinct from the detection) of a resolved magnetic moment to measure *both* s_1 and s_2.

A more justifiable estimate of the resolved magnetic moment was obtained from the deflection pattern of two just visibly resolved components obtained by the spontaneous development of an initially invisible trace of only $1\frac{1}{2}$ hr. exposure. *Interpreted as simple*, the widths of the deflected traces, compared with H, gave $Mg = \pm \frac{1}{3}$.

The experimental error cannot possibly have been less than 10 per cent.; hence the fact that the traces, *interpreted as simple*, gave $Mg = \frac{1}{3}$ does *not* exclude the possibility that states other than $^2P_{\frac{1}{2}}$ may have been present in very considerable amount. (Compare, for example, the case of bismuth, where two components of $\mu_1 = 0.72\mu_B$, $\mu_2 = 2.16\mu_B$ present in equal amounts, combined to give a summation curve which interpreted as simple gave $\mu_i = 0.85\mu_B$.) All that can be said from the experimental results is that the $^2P_{\frac{1}{2}}$ state is present in active nitrogen to an undetermined, not necessarily predominant extent.

It was observed that the presence of the yellow after-glow in the discharge tube was not essential to the production of a beam of active nitrogen; this confirms the observation of other workers that the chemical activity of active nitrogen may persist in the non-glowing state.

Jackson and Powell have recently developed a molecular ray method for the study of active nitrogen which promises to become an important general method for the study of metastable states. Although not a magnetic method, it may conveniently be discussed here.

The essence of the method is the measurement of the energy of the photo-electrons which are liberated from a metallic target when metastable atoms or molecules impinge upon it and in so doing suffer transitions to a lower energy level. In the experiments of Jackson and Powell, a beam of active nitrogen of circular cross section impinges on an outgassed nickel target, which is surrounded by a collecting cylinder connected with a Compton electrometer. The electron current between target and collector is plotted against a negative stopping potential applied to the collector.

The work function ϕ of nickel is 4 volts; hence with this target the molecular $^3\Sigma$ state (8 volts) should make itself evident, but not the atomic ^2P (3.5 volts) and ^2D (2.3 volts) states. The preliminary experiments brilliantly confirm the spectroscopic data. A definite kink on the current: stopping potential curve does in fact occur with the collector at -4 volts against the target ($8\ (^3\Sigma) - 4\ (\phi_{Ni}) = 4$ volts). There is interesting evidence of fine structure of the $^3\Sigma$ level, which is being further investigated.[a]

[a] I am indebted to Dr Jackson for the above information. For a preliminary note, see *B. A. Report*, 1930, p. 295.

NUCLEAR MOMENTS

The individual components of a spectral multiplet themselves often possess a so-called "hyperfine structure" (H.F.S.), which cannot be deduced from the motions of a configuration of electrons having the properties of charge, mass, and spin. It is therefore natural to look for the origin of H.F.S. in the nucleus; and it has in fact been found possible to ascribe the observed cases of H.F.S. to one or other nuclear property.

Now by far the commonest type of H.F.S. arises from the presence of a nuclear *magnetic* moment. The difficulties of spectroscopic observation are, however, so great that up to the present it has not been found possible to do more than demonstrate the existence of nuclear moments of the expected order of magnitude, $10^{-3} \mu_B$.[a] It is therefore important to see how far the molecular ray method is capable of being developed to measure moments of this order; it being remembered at the outset that the method is restricted in any event, with the resolution so far obtained, to atoms or molecules (such as Zn, Cd, Hg, etc.; H_2O, H_2, etc.) which do not possess an electronic magnetic moment.

In the first place, it is clear that inhomogeneities far in excess of the 10^4 gauss/cm. sufficient for the determination of electronic moments are necessary. Attempts to detect, far less measure, nuclear magnetic moments from a broadening of the trace in a normal Stern-Gerlach set-up, such as have been made with I_2 by Rodebush and Nichols[b], are, because of the inherently poor definition of the boundary of the penumbra, lateral motion, etc., unlikely to succeed. A well-defined deflection must be aimed at. With slit widths of the order 0·01 mm., the deflection should be at least 0·01 mm.; with a total path of 20 cm. this requires an inhomogeneity of at least

[a] See Pauling and Goudsmit, *The Structure of Line Spectra*, Chapter XI, for an account of the work on H.F.S. up to the end of 1929.

[b] Rodebush and Nichols, *J. Amer. Chem. Soc.* **52**, 3024, 1930.

10^6 gauss/cm. Such a high inhomogeneity can however be maintained in a region whose cross section is at the most $0\cdot1 \times 0\cdot1$ mm., which immediately sets a limit $h \sim 0\cdot1$ mm. to the slit length. Since $I \propto h/r^2$ (cf. p. 25), the intensity is thereby so drastically diminished as to be experimentally useless.

This difficulty can be overcome by using a technical device due to Stern,[a] which he calls a multiplier ("Multiplikator"). Suppose that the high inhomogeneity requires the channel in the channelled pole-piece to be so narrow that the intensity falls to $1/n$th of the smallest detectable value. Then the intensity can be raised again to the required value by cutting in the pole-piece not one, but n channels converging to a

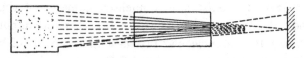

Fig. 53

common focus at the position of the detector, so that the individual traces overlap. Calculation shows that by this means moments of the order $10^{-5}\mu_B$ should still be detectable.

The arrangement as devised experimentally by Knauer and Stern[b] for the detection of moments of the order $10^{-3}\mu_B$ is shown schematically in Fig. 53. The oven slit lies in the plane of the multiplier, which is furnished with 150 channels, each $0\cdot1$ mm. wide and of about the same depth, cut with a diamond on the plane face of an iron pole-piece. Over against it, some 4 to 13μ above the plane of the paper, must be pictured the second pole-piece, which is flat. The inhomogeneity (ca. 10^6 gauss/cm.) in the plane of the channel entrances must of course be calculated, the values found being checked against the direct measurements made with single large channels. The image slit in Knauer and Stern's experiments was placed

[a] Stern, *Z. Physik*, **39**, 757, 1926 (U. z. M. 1).
[b] Knauer and Stern, *Z. Physik*, **39**, 780, 1926 (U. z. M. 3).

for convenience of adjustment *after* the field.[a] The deflections of the focal spot are at right angles to the plane of the paper, towards and away from the reader.

The arrangement is not ideal, in so far as it is possible for molecules to fly diagonally across the multiplier, as indicated by the broken line in the figure. Such molecules in their flight pass alternately over the channel and over the interspaces; thus the direction of the deflecting force is ever and again changing its direction, so that they are deflected now up, now down, and hence on the average are not deflected at all. Thus only at the focus of the multiplier can deflected molecules be expected.

The magnetic moment of the water molecule, of the order $10^{-3}\mu_B$, arising from the rotation of the H nuclei about the oxygen ion, was actually detected by Knauer and Stern. Plate V, Fig. 54, is a photograph, intensified by reprinting, of the trace with field; the intensity at the position of the focus is clearly diminished, which shows that a fraction of the molecules has been deflected at the expected point. The absence of a definite deflection pattern is exactly what one should expect. Apart from the uncertainty in the path of the molecules over the multiplier, discussed in the preceding paragraph, it is clear that the velocity distribution of the molecules in the beam, and the presence of many rotation states, must give rise to a diffuse intensity distribution, the details of which could not be expected to emerge from the target background.

[a] If l_1 is the distance from source slit to entrance of field, l_2 the length of the field, l_3 the distance from the exit of the field, where the image slit is placed, to the detector; then the deflection s suffered by a magnetic dipole, mass m, resolved magnetic moment μ_i, velocity v, is

$$s = \frac{\mu_i}{mv^2} \cdot \frac{\partial H}{\partial z}\left(l_2 l_3 - \tfrac{1}{2}l_2^2 \cdot \frac{l_3}{l_1 + l_2}\right).$$

Very little if anything is gained in resolving power by placing the image slit after the field; it is true that the traces are thereby narrower; on the other hand, the deflections are in an almost corresponding degree smaller.

The multiplier technique has not been followed up since the original orientating experiments of Knauer and Stern; but it seems clear that the device could be perfected, above all by the elimination of cross-fire, to be capable not merely of detecting, but also of measuring nuclear magnetic moments.

ABSOLUTE MEASUREMENTS

Two notable advances in the technique of the deflection method were made in 1929. The first is due to Rabi,[a] who showed that it was possible to replace the exceedingly difficult

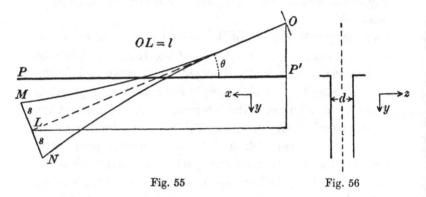

Fig. 55   Fig. 56

measurements of the inhomogeneity from point to point of the field by the determination of field strengths only; the second is the application by Taylor[b] of Langmuir and Kingdon's surface ionisation gauge to the quantitative measurement of alkali metal beams. These developments taken together make possible immediately the absolute determination of resolved magnetic moments correct to about 1 in 500. This degree of accuracy, while admittedly falling short of that needed for the study of certain important theoretical points, is nevertheless great enough to justify a brief survey of the experimental possibilities.

[a] Rabi, *Z. Physik*, **54**, 190, 1929 (U. z. M. 12).
[b] Taylor, *Z. Physik*, **57**, 242, 1929 (U. z. M. 14).

We begin with a strict analysis of the Rabi method, which was sketched on pp. 117–19. Fig. 55 represents a side elevation of the arrangement in the xy plane (cross section, Fig. 56). OL is the parent beam, emerging from the image slit O, which is taken to be the origin of co-ordinates. MLN is a target placed perpendicularly to OL. PP' is the geometrical boundary of the field. For simplicity we consider a beam of atoms for which $J = \frac{1}{2}$, giving two deflected components OM, ON, which are to a high degree of approximation symmetrical with reference to OL.

The equations of motion in the xy plane of an atom of mass m, resolved magnetic moment μ_i, are

$$\left.\begin{aligned} m\ddot{x} &= 0 \\ m\ddot{y} &= \mu_i \frac{\partial H}{\partial y} \end{aligned}\right\} \qquad \ldots\ldots(5\cdot21),$$

whence, if E_0 is the energy of translation of the atom at the origin,

and

$$\left.\begin{aligned} \dot{x} &= x_0 = \sqrt{\frac{2E_0 \cos^2 \theta}{m}} \\ \dot{y} &= \sqrt{\frac{2E_0 \sin^2 \theta}{m}} (1 + \gamma) \end{aligned}\right\} \qquad \ldots\ldots(5\cdot22).$$

where

$$\gamma = \frac{\mu_i (H - H_0)}{E_0 \sin^2 \theta}$$

Hence

$$\left.\begin{aligned} \int_0^{y_1} (1 + \gamma)^{-\frac{1}{2}} \, dy &= \int_0^{x_1} dx \tan \theta \\ x_1 &= l \cos \theta - s \sin \theta \\ y_1 &= l \sin \theta + s \cos \theta \end{aligned}\right\} \qquad \ldots\ldots(5\cdot23),$$

where

whence, neglecting powers of γ higher than the second,

$$s = \tfrac{1}{2} \cos \theta \int_0^{y_1} (\gamma - \tfrac{1}{4}\gamma^2) \, dy \qquad \ldots\ldots(5\cdot24).$$

A first approximation gives

$$s_0 = \frac{\mu_i (\overline{H - H_0}) l}{2E_0 \tan \theta} \qquad \ldots\ldots(5\cdot25),$$

where the field strength is averaged in the y direction over the distance $y' = l \sin \theta$.

To a second approximation

$$\left.\begin{array}{c} s = s_0 \left(1 + k s_0\right) \\ k = \dfrac{1}{l \tan \theta} \left[\dfrac{(H - H_0)}{(\overline{H - H_0})} - \dfrac{\overline{(H - H_0)^2}}{2 \, (\overline{H - H_0})^2} \cdot \dfrac{1}{\cos^2 \theta}\right] \end{array}\right\}$$

......(5·26).

If, in (5·25), $(\overline{H - H_0}) = 10^4$ gauss; $l = 10$ cm.; $\theta = 10°$; and if E_0 is taken to be the most probable kinetic energy at $1000°$ K.: then since $\mu_i = 0.918 \times 10^{-20}$ erg gauss^{-1}, $s_0 = 400\,\mu$. It is then evident from (5·26) that the second approximation suffices if the accuracy aimed at is not higher than about 1 in 500.

The above analysis refers to the plane of symmetry only; an actual beam has a finite height, in the z direction. It can readily be shown, however, that over a height of about $0.8d$, neglect of the force $\mu_i \dfrac{\partial H}{\partial z}$ introduces an error in the value of s of less than 1 in 1000; in other words the deflected traces are practically straight. Plate V, Fig. 50, which is due to Rabi, shows how well this is borne out under the actual experimental conditions.

Ability to measure the intensity distribution, both in the parent beam and in the deflected components, makes possible a rigorous evaluation of μ_i; for from the measured intensity distribution in the parent beam one can calculate for an arbitrary μ_i the intensity distribution in a deflected component, and then by trial and error find the particular value of μ_i for which the calculated intensity distribution in the deflected beam coincides with that measured experimentally.

The intensity at any point of an alkali metal beam can be measured with the surface ionisation detector correct to 1 in 10,000 of the intensity in the umbra of the parent beam. Fig. 57 shows the plot of a potassium deflection pattern made by Taylor with the surface ionisation detector. The remanent intensity which is observable at the position of the undeflected beam arises partly from the presence of undeflected molecules

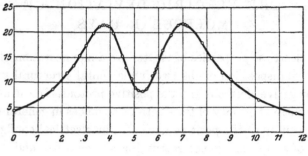

Fig. 57

(see below, p. 189) and partly from incipient cloud formation at the oven slit (see above, p. 18); both effects can be eliminated by careful choice of the oven conditions and the lay-out of the slit system.

Chapter 6

THE ELECTRIC DEVIATION OF MOLECULAR RAYS

The electrical properties of matter are a manifestation of the relative disposition of the two fundamental structural units, the negative electrons and the positive protons. In the atom, the centre of gravity of the electrons coincides in general with the position of the positive nucleus. The electrical configuration of a molecule is necessarily less symmetrical, and it frequently happens that the centre of gravity of the electronic configuration does not coincide with that of the nuclei. In such case, the molecule possesses a permanent electric dipole

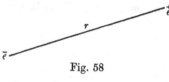

Fig. 58

(see Fig. 58); the dipole moment of the molecule is then $\mu_e = e \cdot \mathbf{r}$, the direction of the moment being reckoned from the negative charge. It is easy to get a rough estimate of the magnitude of μ_e: e will be of the order of the electronic charge, $4 \cdot 77 \times 10^{-10}$ E.S.U., r will be of the order of a molecular diameter, 10^{-8} cm.; hence $\mu_e \sim 10^{-18}$ E.S.U.

The measurement of the electric dipoles of molecules is clearly capable of yielding much information about the arrangement of their component atoms. The standard method of measurement at present is that due to Debye,[a] whereby μ_e is derived from the variation with temperature of the dielectric constant of a substance, whether as a gas, or in dilute solution in a non-polar solvent. The method is subject to the limitations imposed, on the one hand by the considerable gaseous pressures which are necessary for reasonably accurate D.C. measurements, on the other by the solubility of

[a] Cf. Debye, *Polar Molecules*, New York, 1929.

the substance it is desired to examine in suitable non-polar solvents.

Now just as the *magnetic* properties of molecules can be studied by the deviation of a molecular ray in an inhomogeneous *magnetic* field, so in precise analogy can the *electrical* properties be studied by deviation experiments in an inhomogeneous *electric* field. When it is remembered, on the one hand, that oven pressures of at the most 1 mm. are required, on the other, that substances insoluble in non-polar solvents can be studied equally with those which are soluble, it will be realised that the molecular ray method can be usefully employed in corroboration and extension of the dielectric constant measurements. Moreover, as we shall see later, the molecular ray method is capable of giving far more information about the energy states of the molecule than the D.C. method.

MOLECULAR ROTATION

Since $\mu_e \sim 10^{-18}$ E.S.U., and since it is comparatively easy to obtain an electrical inhomogeneity $\frac{\partial X}{\partial s}$ of some 10^4 E.S.U./cm. (for example, near a charged wire), it might appear at first sight that the deflections in an electric deviation experiment might be some hundred times those obtained in the measurement of atomic magnetic moments $\left(\mu_i . \frac{\partial H}{\partial s} \sim 10^{-20} . 10^4 \sim 10^{-16} \right)$. This is however not generally the case, on account of the *temperature rotation* of the molecules. We shall make this clear with reference to a simple model of a diatomic molecule, considered as a rigid dumb-bell, of which HCl may be taken as the type; remembering of course that the conclusions arrived at are not necessarily applicable to polyatomic molecules.

Fig. 59 illustrates the pure rotation of such a rigid model of HCl about an axis AB through its centre of gravity. Since the direction of the dipole moment μ_e coincides with that of the dumb-bell, the effect of the rotation is clearly this: the externally observable dipole moment, averaged over the period of a complete rotation, vanishes. The presence of an electric field however distorts the uniform rotation of the dipole, so that there results a time-averaged moment $\bar{\mu}$, proportional to the field strength, which, if the electric field be inhomogeneous, is acted on by a force $\bar{F} = |\bar{\mu}| \cdot \dfrac{\partial X}{\partial s}$. Thus a beam of these molecules when shot through an inhomogeneous electric field suffers a deviation, from which, if the relation of

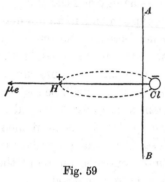

Fig. 59

$\bar{\mu}$ to μ_e is known, the electric moment of the molecules can be ascertained.[a]

CALCULATION OF THE TIME-AVERAGED MOMENT

We proceed to the evaluation of $\bar{\mu}$. The calculation was first carried out by Kallmann and Reiche.[b] Their analysis is however very cumbersome, and we shall adopt here a simple method due to Stern.[c] It is assumed at the outset *that the temperature of the experiment is high,* so that such a multitude

---

[a] $\bar{\mu}$ should not be confused with the *polarisability a.*

a is defined as the electric moment induced by unit field; it arises from the deformability of the electronic configuration of the molecule, and the sense of a is thus *always* in the direction of the field. All molecules are polarisable to a greater or less extent, whether they possess a dipole moment or not.

$\bar{\mu}$, on the other hand, is a *time-averaged moment* arising from the perturbation of the motion *of a dipole* μ_e. The *sense* of $\bar{\mu}$ may be with or against the field (see equation (6·9), p. 159).

In the text, the polarisability a is neglected in deriving $\bar{\mu}$, since the molecule has been assumed rigid.

[b] Kallmann and Reiche, *Z. Physik,* 6, 352, 1921.

[c] I have to thank Professor Stern for his permission to use this demonstration here.

of discrete rotation states of the molecules is present that a classical distribution of rotational energy may be postulated. Two special cases are then considered: (1) when the axis of rotation is perpendicular to the field direction, (2) when the axis of rotation is parallel to the field direction; from these, the result is generalised to include an arbitrary inclination ϕ of the axis of rotation to the field.

First Special Case. Fig. 60. Let us first consider qualitatively the effect of the field on the uniform rotation of the dipole. Starting from an axis of reference at right angles to the field direction, the angular velocity ψ of the molecule when $\psi = 0$ is ω_0, its value in field-free space. In the first quadrant $(0 < \psi < \pi/2)$, the rotation is accelerated, in the second quadrant retarded, until when $\psi = \pi$ the angular velocity regains its initial value ω_0. The angular velocity is a maximum at $\psi = \pi/2$. In the third quadrant the motion is retarded, in the fourth accelerated to its initial value ω_0 at $\psi = 2\pi$. The angular velocity is a minimum at $\psi = 3\pi/2$. The nett result is that the dipole takes longer to describe the last two quadrants than it does to describe the first two; and there results, independently of the direction of rotation of the dipole, a time-averaged moment, whose direction is *opposed* to that of the field, namely

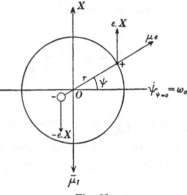

Fig. 60

$$\bar{\mu}_1 = \frac{\int_0^T \mu_e \cdot \sin \psi \cdot dt}{\int_0^T dt} \qquad \ldots\ldots(6 \cdot 1),$$

where T is the period of rotation.

If now I is the moment of inertia of the molecule, the equation of motion in the field X is

$$I\ddot\psi = \mu_e X \cdot \cos\psi \qquad \ldots\ldots(6\cdot2),$$

which expresses the fact that the rate of change of angular momentum of a rigid body is equal to the moment of the external forces.

Integrating, we have

$$\dot\psi = \omega_0 \left(1 + \frac{2\mu_e X}{I\omega_0{}^2} \cdot \sin\psi\right)^{\frac{1}{2}} \qquad \ldots\ldots(6\cdot3).$$

Hence, from (6·1),

$$\bar\mu_1 = \mu_e \cdot \frac{\displaystyle\int_0^{2\pi} \sin\psi \left(1 + \frac{2\mu_e X}{I\omega_0{}^2} \cdot \sin\psi\right)^{-\frac{1}{2}} d\psi}{\displaystyle\int_0^{2\pi} \left(1 + \frac{2\mu_e X}{I\omega_0{}^2} \cdot \sin\psi\right)^{-\frac{1}{2}} d\psi}$$

$$\ldots\ldots(6\cdot4).$$

Since $\dfrac{2\mu_e X}{I\omega_0{}^2}$ is in practice small compared with unity, we need expand the expressions within brackets to the first power of $\sin\psi$ only, and obtain

$$\bar\mu_1 = -\frac{\mu_e{}^2 X}{I\omega_0{}^2} \cdot \frac{\displaystyle\int_0^{2\pi} \sin^2\psi\, d\psi}{\displaystyle\int_0^{2\pi} d\psi}$$

$$= -\frac{\mu_e{}^2 X}{2I\omega_0{}^2} \qquad \ldots\ldots(6\cdot5).$$

Second Special Case. Fig. 61. Here the effect of the field is to cause the dipole axis to tilt through an angle α, in the sense that the dipole tends to align itself with the field. Thus there results in this case a time-averaged moment $\bar\mu_2$ *in the direction of* the field, independently of the direction of rotation of the dipole. The field-free motion of pure rotation is converted into one of precession, and the problem is simply that of the centrifugal governor. We have, since α is small,

$$\alpha = \frac{eX}{m_1 \omega_0{}^2 d_1} = \frac{eX}{m_2 \omega_0{}^2 d_2} \qquad \ldots\ldots(6\cdot6),$$

where m_1, m_2 are the masses, d_1, d_2 the distances from the centre of gravity O, of the positive and negative ends of the dipole.

Again, $I = m_1 d_1 (d_1 + d_2) = m_1 d_1 r = m_2 d_2 r$ ......(6·7).

Hence $\bar{\mu}_2 = e.(d_1 + d_2) \sin \alpha$

$= \mu_e . \alpha$ since α is small

$= \dfrac{\mu_e{}^2 X}{I \omega_0{}^2}$ ......(6·8)

from (6·6) and (6·7).

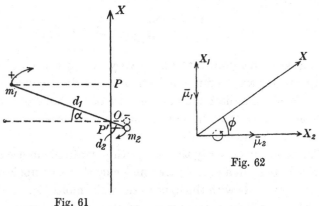

Fig. 61

Fig. 62

Generalising, let us suppose that the axis of rotation makes an angle ϕ with the direction of the field X. Then (Fig. 62)

$\bar{\mu} = \bar{\mu}_1 \sin \phi + \bar{\mu}_2 \cos \phi$

$= -\dfrac{\mu_e{}^2 X}{2 I \omega_0{}^2} . \sin^2 \phi + \dfrac{\mu_e{}^2 X}{I \omega_0{}^2} . \cos^2 \phi$ from (6·5) and (6·8)

$= \dfrac{\mu_e{}^2 X}{2 I \omega_0{}^2} . (3 \cos^2 \phi - 1)$ ......(6·9),

which establishes the final relation between $\bar{\mu}$ and μ_e.

A rough estimate of the order of magnitude of $\bar{\mu}$ for diatomic molecules may be obtained for example from the first special case. Equation (6·5) may be written in the form

$$\bar{\mu} = \mu_e . \frac{\mu_e X}{4 E_r},$$

where E_r is the energy of rotation. Or, remembering that a pure rotator has two degrees of freedom,

$$\bar{\mu} = \mu_e \cdot \frac{\mu_e X}{4kT} \qquad \dots\dots(6\cdot10),$$

in which T is the absolute temperature, and k is Boltzmann's constant, $1\cdot37 \times 10^{-16}$ ergs per degree. If we take the field strength X to be 150,000 volts/cm., that is 500 c.g.s. units, then setting the numerical values in $(6\cdot10)$ we have, at room temperature,

$$\bar{\mu} \sim \mu_e \cdot \frac{10^{-18}.500}{4.1\cdot37 \times 10^{-16}.300} \sim 3 \times 10^{-3}\mu_e.$$

It is seen therefore that unless very strong fields and low temperatures are used the deflections in an electric deviation experiment on a diatomic molecule are of the same order as, or even smaller than those obtained in the measurement of atomic magnetic moments.

This is not necessarily the case with polyatomic molecules. It may happen in a complicated molecule that the dipole axis does not coincide with the figure axis of the molecule. In that case, there will be a component of the dipole moment perpendicular to the figure axis (or what amounts to the same thing, parallel to the axis of rotation). In spite of the fact of rotation therefore there will remain a dipole moment independent of the field strength, and large deflections in the field will result. A merely qualitative observation of the scale of the deflection pattern would then suffice to give a decision in cases where an intermolecular configuration is in doubt.

INTENSITY DISTRIBUTION IN THE
DEFLECTED BEAM

It is easy to construct the electrical analogues of equations (5·4) and (5·6′) or (5·25), by means of which a measurement of the deflection s can be used to determine μ_e. For in the electrical case $F_2 = \bar{\mu} \dfrac{\partial X}{\partial z}$, and from (6·9)

$$
\left.
\begin{aligned}
s &= C \cdot \frac{3 \cos^2 \phi - 1}{E_t E_r} \\[2mm]
C &= \frac{\mu_e^2}{16} \cdot X \cdot \frac{\partial X}{\partial z} \cdot l^2 \\[2mm]
C &= \frac{\mu_e^2}{16} \cdot X \cdot \frac{\partial X}{\partial z} \cdot l_1^2 \left(1 + 2l_2/l_1\right)
\end{aligned}
\right\} \quad \ldots\ldots (6·11)
$$

or

if the detector is placed a distance l_2 behind the end of the field, itself of length l_1.

Equations (6·11) are applicable when X and $\dfrac{\partial X}{\partial z}$ have the same direction (for example, near a charged wire).

It is clear that the Rabi method, discussed for the magnetic case in Chapter 5, is equally applicable to the measurement of electric dipole moments; the beam is here shot at an angle between the plates of a parallel plate condenser. At the present stage of the technique we can hope for an accuracy of not more than a few per cent. in electric deviation experiments, and we need therefore consider only the expression for the deflection in the approximate form of (5·25). We write (5·25) in the general form

$$
s = \frac{1}{2E_t} \cdot \frac{\bar{E}l}{\tan \theta} \qquad \ldots\ldots (6·12),
$$

where $\bar{E}$ is the mean energy of the molecule due to the field, averaged over the path of the undeflected beam. Now the electric case for diatomic molecules differs from the magnetic case inasmuch as we have here to deal with a time-averaged moment $\bar{\mu}$ *proportional to the field strength.* We have $\bar{\mu} = \beta X$; the force on the molecule at any point in the plane of symmetry (xy plane) is $F_y = \beta X \dfrac{\partial X}{\partial y}$, and the energy in the field is

$$E = \int_0^y F_y \, dy = \beta \int_{X_0}^X X dX = \tfrac{1}{2}\beta \, (X^2 - X_0{}^2),$$

whence $\quad \bar{E} = \tfrac{1}{2}\beta \, (\overline{X^2 - X_0{}^2}) \qquad \ldots \ldots (6\cdot 13).$

Substituting the value of β from (6·9) in (6·13), and substituting the resulting value of $\bar{E}$ in (6·12), we have

$$s = C' \cdot \frac{3\cos^2\phi - 1}{E_t E_r}$$

$$\left. \begin{array}{c} \\ \\ \end{array} \right\} \quad \ldots \ldots (6\cdot 14).$$

where $\qquad C' = \dfrac{\mu_e{}^2}{16} \cdot \dfrac{(\overline{X^2 - X_0{}^2}) \, l}{\tan\theta}$

Now (6·11) and (6·14) differ only in the factor C or C'; we concentrate attention therefore on the factor $\dfrac{3\cos^2\phi - 1}{E_t E_r}$.

First, we observe that the sign of $(3\cos^2\phi - 1)$ for a particular molecule decides the sign of its deflection in an inhomogeneous electric field. Since the direction of $\bar{\mu}$ is independent of the direction of rotation of the dipole for a given inclination ϕ, we need for this purpose consider only values of ϕ lying between 0 and $\pi/2$. It is easily seen that

$$s > 0 \text{ when } \tfrac{1}{3} < p^2 \leqslant 1,$$
$$s < 0 \text{ when } 0 \leqslant p^2 < \tfrac{1}{3},$$

where p has been written for $\cos\phi$; s being reckoned positive when in the direction of $\dfrac{\partial X}{\partial z}$ increasing.

Next we note that the intensity distribution in the deflected beam will be exceedingly complicated; for the deflection s depends simultaneously on three quantities: (1) the energy of translation E_t, the values of which are grouped about a most probable value governed by the Maxwell distribution of translational velocities at the temperature of the source; (2) the energy of rotation E_r, whose values at high temperature are governed by the equipartition law; (3) the inclination ϕ of the axis of rotation to the field direction.

In order therefore to calculate the intensity distribution in the deflected beam it is necessary to calculate (1) the probability W_1 that E_t lies between E_t and $E_t + dE_t$; (2) the probability W_2 that E_r lies between E_r and $E_r + dE_r$; (3) the probability W_3 that ϕ lies between ϕ and $\phi + d\phi$. This we proceed to do.

1. The probability of a molecule with a velocity in the source between v and $v + dv$ arriving at the detector is by (2·4)

$$W(v)dv = 2/\alpha^4 . e^{-v^2/\alpha^2} . v^3 . dv,$$

whence, changing the independent variable from v to

$$E_t = E_a . v^2/\alpha^2,$$

we have at once that the probability of molecules with energies between E_t and $E_t + dE_t$ arriving at the detector is

$$W_1(E_t)dE_t = e^{-E_t/E_a} . E_t/E_a^2 . dE_t$$

$$= \frac{1}{(kT)^2} . e^{-E_t/kT} . E_t . dE_t \quad ...(6\cdot15).$$

2. The probability that a molecule arrives at the detector with a rotational energy between E_r and $E_r + dE_r$ is just the probability of the existence of such molecules in the source; which for a pure rotator is

$$W_2(E_r)dE_r = 1/kT . e^{-E_r/kT} . dE_r \quad ......(6\cdot16).$$

3. The probability that ϕ lies between ϕ and $\phi + d\phi$ is, since the beam is essentially collision free, derived purely geometrically as the ratio of the area of the zone of a unit sphere between co-latitudes ϕ and $\phi + d\phi$, to the total area of the sphere: or since only the range of values $0 < \phi < \pi/2$ need here be considered, to half the area of the sphere. That is

$$W_3(\phi)d\phi = \frac{2\pi \sin \phi \, d\phi}{2\pi} = \sin \phi \, d\phi \quad ...(6\cdot17).$$

Let us now define a new quantity σ, such that

$$\left.\begin{array}{l} \sigma = s/s_0 \\ s_0 = 2C/(kT)^2 = 2C'/(kT)^2 \end{array}\right\} \quad \ldots\ldots(6\cdot18),$$

s_0 being the deflection corresponding to $\phi = 0$ and the most probable velocity α. Further, let

$$E_t/kT = y.$$

E_r can now be expressed in terms of σ. For

$$\left.\begin{array}{l} \sigma = s/s_0 = \dfrac{kT}{2} \cdot \dfrac{3p^2 - 1}{E_r y} \\[2ex] E_r = \dfrac{kT}{2} \cdot \dfrac{3p^2 - 1}{\sigma y} \\[2ex] dE_r = -\dfrac{kT}{2} \cdot \dfrac{3p^2 - 1}{\sigma^2 y} \cdot d\sigma \end{array}\right\} \quad \ldots\ldots(6\cdot19).$$

whence

and

Hence finally the probability $W(\sigma)\,d\sigma$ that σ lies between σ and $\sigma + d\sigma$ is, using equations (6·15) to (6·19),

$$\int_{p_1}^{p_2} dp \int_0^\infty e^{-\frac{3p^2-1}{2\sigma y}} \frac{3p^2 - 1}{2\sigma^2 y}\, d\sigma \cdot e^{-y} y\, dy,$$

where

$$p_1 = 1; \ p_2 = 1/\sqrt{3} \text{ when } \sigma > 0,$$

$$p_1 = 0; \ p_2 = 1/\sqrt{3} \text{ when } \sigma < 0.$$

The expressions just found can only be integrated numerically. The rather laborious calculations have been carried out by Feierabend;[a] his results are embodied in Table IX, and are exhibited graphically in Fig. 63.

We may note the following features of the intensity curve of Fig. 63. (1) The maximum of the curve is very close to the position of the parent beam. (2) The molecules are deflected both right and left of the position of the parent beam. (3) The average moment of the attracted molecules (σ positive) is greater than that of the repelled molecules (σ negative).

It must not be forgotten that the curve of Fig. 63 is an ideal curve inasmuch as the polarisability α of the molecules has

[a] Hamburger Staatsexamensarbeit (unpublished).

Table IX

σ	$W(\sigma)$	σ	$W(\sigma)$	σ	$W(\sigma)$	σ	$W(\sigma)$
− 5	0·0065	− 0·4	0·353	− 0·06	1·394	0·16	0·562
− 2	0·0292	− 0·3	0·476	− 0·05	1·436	0·25	0·393
− 1·5	0·0504	− 0·2	0·735	− 0·04	1·442	0·50	0·190
− 1·0	0·0966	− 0·18	0·808	− 0·03	1·423	1·00	0·0835
− 0·9	0·114	− 0·16	0·898	− 0·02	1·350	1·50	0·0327
− 0·8	0·135	− 0·14	0·994	− 0·01	1·247	2·00	0·0265
− 0·7	0·163	− 0·12	1·087	0·00	1·153	3·00	0·0142
− 0·6	0·204	− 0·10	1·190	0·05	0·901	3·50	0·0104
− 0·5	0·263	− 0·08	1·309	0·10	0·732	4·00	0·0081

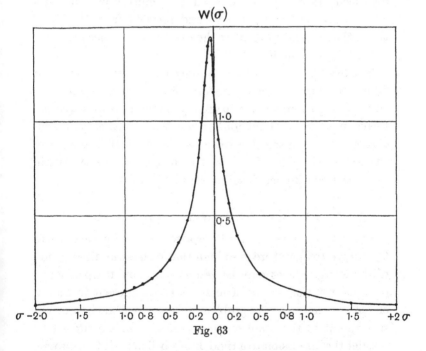

Fig. 63

been neglected (cf. footnote, p. 156). The general effect of the polarisability is to shift the curve bodily in the positive direction. To take accurate account of it is rather difficult; at present it would seem best from the quantitative side to concentrate attention on substances for which α is comparatively small, so that the simplified theory may first be firmly established.

The general features of the curve of Fig. 63 have been verified by Wrede,[a] using KI. The beam was shot parallel to a charged wire, and at a short distance from it. Plate VI, Fig. 64 is a KI trace at different stages of dry development. The central portion shows the influence of the inhomogeneous electric field around the wire: first, the whole of the central portion has been displaced towards the wire, due to the polarisability; second, on either side of the maximum of the displaced trace can be detected the imprint of molecules which have been attracted to and repelled from the wire; third, the average moment of the repelled molecules is clearly comparatively small.

Wrede's results are of course only qualitative; indeed, an accurate determination of the form of the intensity-deflection curve at high temperature can only be made with a quantitative method of detection. Even if the beam be monochromatised by a suitable velocity selector, little is gained in the way of simplification, for at high temperature there still remains a continuous distribution of E_r and ϕ.

THE CASE OF LOW TEMPERATURE

Hitherto we have confined our attention to the case where the source temperature is so high that a classical distribution of the energy states may be assumed. At low temperatures, only the rotation states of low energy are present in appreciable amount, and the problem must be regarded from the standpoint of the quantum mechanics.[b] Once again let us restrict the discussion to a rigid dumb-bell model of a diatomic molecule; then the angular momentum in field-free space is

$$J^* = [J\,(J+1)]^{\frac{1}{2}} \cdot h/2\pi \qquad \ldots\ldots(6\cdot20),$$

[a] Wrede, Z. Physik, 44, 261, 1927 (U. z. M. 7).

[b] Mensing and Pauli, Physikal. Z. 27, 509, 1926; Kronig, Proc. Nat. Acad. Sci. 12, 488, 1926.

where $J = 0, 1, 2, ...$ In an external electric field $\mathbf{X}$ the vector $\mathbf{J}^*$ can take up only certain prescribed orientations ϕ relative to the field direction such that $\cos \phi = M/J$; M being defined by the relations

$$- J \rightleftharpoons M \rightleftharpoons J.$$

The energy of rotation is

$$E_J = \frac{h^2}{8\pi^2 I} . J (J + 1) \qquad ......(6.21)$$

and

$$\bar{\mu} = \gamma . \beta (J, M) X$$

where $\quad \gamma = \dfrac{8\pi^2 I \mu_e{}^2}{h^2}$

and

$$\beta (J, M) = \frac{1}{(2J - 1) (2J + 3)} \left(\frac{3M^2}{J (J + 1)} - 1\right) \text{if } J \neq 0$$

or $\qquad = \tfrac{1}{3} \qquad\qquad\qquad\qquad\qquad \text{if } J = 0$

$\left.\begin{array}{c} \\ \\ \\ \\ \\ \end{array}\right\}$(6.22).

The deflection $s_{J, M}$ for a molecule in the state J, M is therefore

$$s_{J, M} = \frac{1}{4E_t} . \gamma . \beta (J, M) . l^2 . X . \frac{\partial X}{\partial z} \quad ......(6.23)$$

for the case that $\mathbf{X}$, $\dfrac{\partial X}{\partial z}$ have the same direction; or for the Rabi method

$$s_{J, M} = \frac{1}{4E_t} . \gamma . \beta (J, M) . l \frac{(X^2 - X_0{}^2)}{\tan \theta} \quad ...(6.24).$$

Now the probability that a molecule is in the state J, M is

$$W_{J, M} = \frac{e^{-E_J/kT}}{\Sigma_{J, M} e^{-E_J/kT}} \qquad ......(6.25),$$

the summation extending over all states.

The theoretical intensity distribution in the deflected beam for a given low temperature T of the source can be found to an approximation sufficiently exact to exhibit its general features, by plotting the intensity curves for the individual states, properly weighted in accordance with (6.25), by means of (5·14), and constructing their summation curve. Fig. 65 shows such a curve—once again an ideal curve in so far as the polarisability of the molecules has been assumed negligible —constructed for the case of HCl at 120° K., which is the lowest source temperature that can be used with a slit of 0·01 mm. if condensation of the gas in the slit is to be avoided.

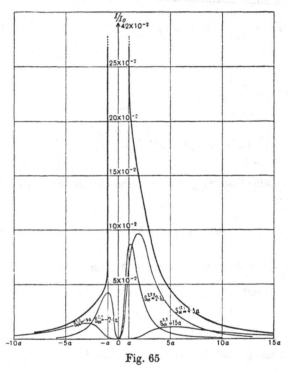

Fig. 65

It should be compared with the high temperature curve of Fig. 63.

It is seen that even for a source temperature as low as 120° K. the traces of the lowest three rotation states $J = 0, 1, 2$ which alone contribute appreciably to the deflected portions of the curve, overlap to such an extent that the resulting summation curve shows only a single maximum, approximately at the position of the undeflected trace.

Thus without the use of a velocity selector, an electric deviation experiment can yield even in very favourable cases only *verification* of the existence of theoretically predicted states. Even allowing for this restriction on its scope, however the molecular ray method is capable of detecting the existence of states which escape the dielectric constant method; for it has been shown, for the case of non-gyroscopic diatomic molecules, that the contribution of the permanent dipole moment to the dielectric constant arises *entirely* from molecules in the lowest energy state $J = 0$. Now it will be seen from Fig. 65 that the state $J = 0$ gives rise to *attracted* molecules only; the repelled molecules which can be detected in Plate VI, Fig. 64, arise therefore from states with $J > 0$, which contribute nothing to the observed dielectric constant.

SCOPE OF THE DEFLECTION METHOD

The experimental material on the method of electrical deviation of molecular rays is extremely meagre, and that which exists is qualitative only. This has tended to give the false impression that the molecular ray method is inherently incapable of giving more than semi-quantitative information about electric dipoles. That is emphatically not the case. It is a question rather of the difficulty of interpreting the results, a difficulty which has been until recently rendered acute by the fact that the target detector cannot deal adequately with a *continuous* intensity distribution. There is an urgent need for accurate work, with a quantitative method of detection, on a standard substance whose characteristics are already well known from the results of other methods. Only in this way can the conclusions of the preceding articles be verified, and made the sound basis for future investigation of untried substances.

THE FIELD

Since for quantitative measurements of electric dipole moments a point-to-point plot of the intensity in the deflected beam is in general *essential*, the characteristics of the inhomogeneous field causing the deflection have to be very carefully considered, even more carefully than in the simpler magnetic cases, when a determination of the *position* of a deflected component often sufficed.

In the first place, a wedge and channel condenser—the electrical analogue of the Stern-Gerlach field—while a possible arrangement, is unsuitable. The reasons are that such an electric field is not merely not calculable, but practically impossible to plot experimentally. Recourse was had therefore in the earlier work to a cylindrical condenser, which is calculable. The condenser takes the form of a thin wire in the axis of a metal cylinder; near the wire the field is sufficiently inhomogeneous to give easily observable deflections.

In Stern's laboratory, the beam was shot parallel to the wire, and a short distance from it. All the deflected traces reproduced in Plate VI were obtained with this field arrangement. They are similar in general appearance to the magnetically deflected traces obtained with the Stern-Gerlach set-up (compare Plate IV, Fig. 47). Unfortunately one is obliged to aim the beam quite close to the wire in order to obtain the requisite inhomogeneity, so that a distorted deflection pattern is with this arrangement unavoidable. The method has another serious disadvantage, namely that the field causing the deflection varies as the inverse cube of the distance from the axis of the cylinder; consequently the attracted molecules are unduly prominent in the deflection pattern.

Clark[a] used a rather different arrangement, necessitated by the poor adjustment possible with his very long beams (over

[a] Clark, *Proc. Roy. Soc.* **124**, 689, 1929.

a metre). The wire, instead of being parallel to the undeflected
beam, was placed *in the beam* and perpendicular to its direction. Molecules passing on either side of it, through windows
cut in the cylinder, would then be deflected. This arrangement
has serious disadvantages: not only does the deflecting force
vary very rapidly close to the wire, but the deflection patterns
obtained from molecules which pass right or left of the wire
respectively must overlap.[a]

The field arrangement which will undoubtedly be used in
all future electric deviation experiments, except perhaps in
some special cases (see footnote, p. 121), is the Rabi field. It
avoids all the disadvantages of the former arrangements. It
has been described in detail for the magnetic case in Chapter 5,
and we shall here merely emphasise those features which are
important for the electric experiments.

The necessary modifications in the equations of Chapter 5
have already been discussed on p. 162. It only remains therefore to evaluate the factor $(\overline{X^2 - X_0^2})$ in (6·14) from the
theory of the parallel plate condenser, the field at the entrance
to which is seen in Fig. 66. If it be assumed that $X_0 = 0$ at
the image slit, we have to evaluate simply $\overline{X^2}$ in terms of the
field between the plates, X_c; it is then easy to show by
graphical methods that

$$\overline{X^2} = 0·6X_c^2.$$

It can be shown further that 90 per cent. of the total change
in $\overline{X^2}$ occurs in a range from $y = d$ to $-y = d/2$; in this respect,
the electrical arrangement is far more favourable than the
magnetic (cf. p. 119).

---

[a] Clark (*loc. cit.*) gives a theory of his method, in which he neglects the
fact of temperature rotation by assuming that the dipole can only orient
parallel and anti-parallel to the field direction. He obtained nevertheless
a double trace of As_2O_3 on the target, in qualitative agreement with his
incorrect equations. In the absence of fuller experimental details, and of a
photograph of the deflected traces, it is impossible to give a meaning to his
results.

Now it has already been pointed out (p. 119) that the extent of this region of inhomogeneity at the field entrance sets a lower limit to the angle θ for a condenser of given length l and distance between the plates d. Geometrical considerations show that for a practicable condenser, for example one with

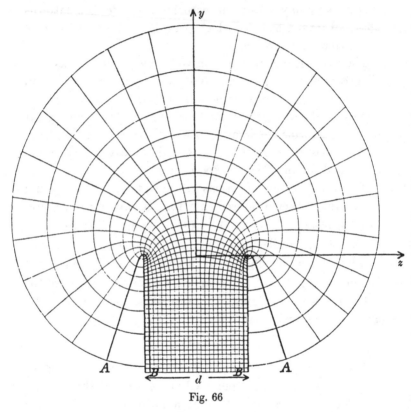

Fig. 66

$l = 10$ cm., $d = 2$ mm., the minimum glancing angle is some $2°$. A numerical value may now be given to $s_0 = 2C'/(kT)^2$ (equations (6·18) and (6·14)) under these conditions. Set for example $\mu_e = 10^{-18}$ E.S.U., $X_c = 500$ E.S.U., $T = 300°$ K.; then $s_0 = 3·6 \times 10^{-2}$ mm.

Trouble from sparking at the edges of the plates when a field of 500 E.S.U. is imposed between them may be avoided

Plate VI

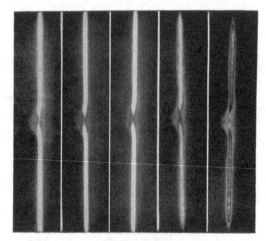

Fig. 64

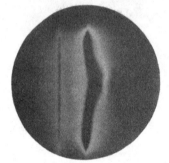

Fig. 67

Fig. 68

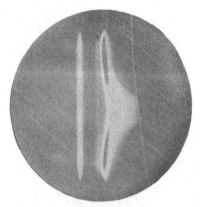

Fig. 69

Electric Deflection Patterns

by rounding the edges so as to follow as far as possible an equipotential (compare the heavy lines AB in Fig. 66). Sharp corners are thus avoided, and at the same time the field at the entrance to the condenser remains exactly as in Fig. 66, and hence calculable.

COMPARATIVE MEASUREMENTS WITH ORGANIC SUBSTANCES

Some interesting *comparative* measurements of the moments of certain organic substances have been made by Estermann[a] with the apparatus of Fig. 12, Chapter 1. It is clear that under very similar experimental conditions (equal field strength and inhomogeneity, equal beam dimensions, also as far as possible equal molecular speeds) the breadth of the deflected trace must give a rough measure of the dipole moment. It is therefore possible in this way to compare the moments of chemically similar substances, for example isomers.

Estermann examined in the first instance certain groups of substances for which the dipole moments were already known from measurement of the dielectric constant. His results are set out in Table X. It is seen that the width of the deflected trace does in fact increase with increasing μ_e, although a visual estimation of the width is too uncertain to reveal the dependance of the deflection on the square of the dipole moment given by (6·11). Insertion of the constants of the apparatus in (6·11) showed that the absolute values of the width are in accordance with a moment of the order 10^{-18} E.S.U.[b]

[a] Estermann, *Z. physikal. Chem.* B, 1, 161, 1928; *ibid.* 2, 287, 1929; *Leipziger Vorträge*, 1929, pp. 17 ff., where full references to the literature are given.

[b] In comparing these values with that calculated for s_0 on p. 172, it must be noted that in Estermann's experiments (1) the length of path from field entrance to detector was very long (14 cm.); (2) the beam was shot very near the charged wire, as is evident from the form of the traces (cf. Plate VI, Figs. 67 and 69).

Table X

Substance	Lateral deviation	Maximum width	Moment from D.C. measurements
	(mm.)	(mm.)	
Diphenyl	ca. 0·1	0·15	0
Diphenylmethane	,, 0·1	0·15	0.4×10^{-18} E.S.U.
Diphenyl ether	,, 0·1	ca. 0·30	1·0
Benzophenone	,, 0·15	,, 0·60	2·5
Methyl ester of			
o-amino benzoic acid	,, 0·1	,, 0·25	1·0
Meta-ester	,, 0·1	,, 0·5	2·4
Para-ester	,, 0·1	,, 0·7	3·3

Estermann, and Estermann and Wohlwill have extended the comparative method to the interesting group of substances Ca_4; that is, methane derivatives with four similar substituents. Penta-erythritol $C(CH_2OH)_4$, which is inaccessible to the D.C. method on account of its insolubility in non-polar solvents, gave the deflection pattern of Plate VI, Fig. 67; to the left is a trace without field. The moment, estimated by comparison with the values of Table X, is 2×10^{-18} E.S.U.[a] Penta-erythritol tetrabromide (Fig. 68) has no permanent moment; the trace shows only an attraction arising from the polarisability. The tetra-acetate, on the other hand (Fig. 69), possesses a large moment. This is in agreement with the D.C. measurements of Ebert and his collaborators on these two substances. Whether the presence of a dipole moment in certain members of the type Ca_4 argues a pyramidal symmetry for the central carbon atom, in accordance with the theory of Mark and Weissenberg, is of course doubtful. All that can be concluded from the experiments is that the *molecule as a whole* does not exhibit tetrahedral symmetry. The question might possibly be settled by an electron diffraction experiment on a Ca_4 vapour beam.

[a] Incidentally, the deflected trace shows very beautifully the presence of molecules which have been repelled from the wire.

Chapter 7

CHEMICAL EQUILIBRIA. IONISATION. SPECTROSCOPIC APPLICATIONS

MOLECULAR EFFUSION

We have seen (Chapter 1) that under conditions of molecular flow through an aperture, the number N of molecules issuing from an aperture of area a in time t is given by the expression

$$N = \frac{N_0 pat}{\sqrt{2\pi RMT}} \qquad \ldots\ldots(7\cdot1),$$

where p is the source pressure in dynes/cm.[2], and the other symbols have their usual significance.

Otherwise expressed, the mass of gas or vapour g in grams effusing in time t is

$$g = pat \sqrt{\frac{M}{2\pi RT}} \qquad \ldots\ldots(7\cdot2).$$

These equations are applicable only so long as there is no *dissociation* of the effusing vapour at the temperature T. In the simple case of the dissociation of diatomic molecules ($7\cdot1$) becomes

$$N_D = N_{X_2} + \tfrac{1}{2}N_X = \frac{N_0 at}{\sqrt{2\pi RMT_D}} \cdot (p_2 + \tfrac{1}{2}\sqrt{2}p_1)$$
$$\ldots\ldots(7\cdot3),$$

where the suffix D refers to the fact of dissociation. p_2 is the partial pressure of the molecules X_2, molecular weight M; p_1 that of the atoms X. The number N_D of particles is supposed measured by a flowmeter *before* dissociation of the gas in the heated region near the orifice.

Equation ($7\cdot2$) becomes

$$g = \sqrt{\frac{M}{2\pi RT}} \cdot at \cdot (p_2 + \tfrac{1}{2}\sqrt{2}p_1) \quad \ldots\ldots(7\cdot4\,a)$$

or

$$g = \sqrt{\frac{at.wt}{2\pi RT}} \cdot at \cdot (p_1 + \sqrt{2}p_2) \quad \ldots\ldots(7\cdot4\,b)$$

according as the molecular weight of the vapour is referred to the molecules X_2 or the atoms X.

Vapour Pressure Determinations. If the molecular weight M of the substance is known, and provided there is no dissociation (or association) over the temperature range to be studied, (7·2) can be very conveniently used to determine the vapour pressures of for example metallic substances, over a pressure range of say 10^{-5} mm. to 1 mm. The method was first suggested by Knudsen.[a]

Substances whose vapour pressures have been determined by the effusion method are, so far as they are known to me, collected in Table XI.

Table XI

Substance	Observer
Li	Bogros, *Compt. rend.* **191**, 322; 560, 1930.
Na	Rodebush and de Vries, *J. Amer. Chem. Soc.* **47**, 2488, 1925.
	Edmundson and Egerton, *Proc. Roy. Soc.* **113**, 520, 1927.
K	Edmundson and Egerton, *loc. cit.*
Cu	Harteck, *Z. physikal. Chem.* **134**, 1, 1928.
Ag	Harteck, *loc. cit.*
Au	Harteck, *loc. cit.*
Zn	Egerton, *Phil. Mag.* **33**, 33, 1917.
Cd	Egerton, *loc. cit.*
Hg	Knudsen, *Ann. Physik*, **29**, 179, 1909; Egerton, *loc. cit.*
Ga	Harteck, *loc. cit.*
Sn	Harteck, *loc. cit.*
Pb	Egerton, *Proc. Roy. Soc.* **103**, 469, 1923.

Vapour pressure determinations are of great practical value, not least for the molecular ray technique itself; but undoubtedly the most important use to be made at the present time of accurate vapour pressure data is in establishing the validity of certain very interesting points in the theory of Thermodynamics.[b] It is therefore worth while to examine rather critically the precautions which must be taken before accurate results can be obtained with the effusion method,

[a] Knudsen, *Ann. Physik*, **29**, 179, 1909.
[b] See particularly Schottky, *Physikal. Z.* **22**, 1, 1921; **23**, 9, 488, 1922.

particularly as it is often the most suitable method to use with some of the substances which are of the greatest interest theoretically.

It is clear from (7·4 b) that if there are diatomic molecules present in a metallic vapour, the effusion method measures, not the total pressure P, but, referred to the atomic weight, a pressure

$$p' = p_1 + \sqrt{2}p_2.$$

It is therefore quite essential to know with certainty either (1) that only one molecular species is present under the conditions of the experiment, in which case (7·2) applies directly; or (2) the amount of association. Thus for example Harteck's values for gallium and tin must remain uncertain until the percentage of Ga_2 and Sn_2 present under the conditions of his experiments shall be determined. Again, it has been known for some time that the vapours of the alkali metals are not strictly monatomic at moderate temperatures; it is therefore fortunate (see p. 189 below) that the molecular ray method has now made possible the accurate determination of the degree of association of Na and K, for which particularly reliable determinations of P and p' are available.[a]

The size and form of the aperture need in all cases careful consideration. Thus (7·2) and (7·4) apply strictly to the case of an ideal aperture only; and it is not always easy, particularly with a glass oven, to obtain a sufficiently sharp edge to the orifice. The *size* of the aperture is, in view of the necessity of obtaining a saturated vapour in the source, of considerable importance, particularly where there is any possibility of contamination of the surface of the charge (cf. Chapter 1, p. 12). Thus Harteck's values for gold are, as he himself remarks, probably too low, because of the large ratio (13·5 mm.[2] : 70 mm.[2]) of the areas of the aperture and gold surface. The total area of the multiple apertures which have

---

[a] See Ladenburg and Thiele, Z. *physikal. Chem.* B, **7**, 161, 1930.

sometimes been used to increase the sensitivity of the method at low pressures requires careful checking against the available surface of the charge.

The time of flow may, unless precautions are taken, be subject to error, because the oven takes time to reach the temperature of the experiment, and meanwhile the vapour is streaming out at a slower rate than that under the final steady conditions. The usual procedure is to make the time of observation long compared with the time necessary to heat the oven. Egerton has occasionally adopted the plan of heating the oven with nitrogen in the apparatus, then rapidly evacuating when the desired temperature is reached; the gas is again admitted at the close of a run. Edmundson and Egerton, in their experiments on sodium, used argon in a similar way.

Bogros avoided uncertainty in the determination of the time of flow by measuring, not the total flow, but that through a second (image) aperture, during the accurately measurable time between the opening and closing of a shutter placed before the oven aperture, the oven having been brought to the desired temperature before the shutter was opened. The possibility of error in determining the quantity of substance, which is only a fraction of the total quantity evaporated, is of course thereby increased.[a]

Dissociation. Equations (7·3) and (7·4) can be used, under the conditions appropriate to each, to determine the degree of dissociation α of a gas or vapour.

Thus it is easy to show that if p_{calc} is the pressure calculated from the measured rate of flow N_D by (7·1), on the assumption

---

[a] Bogros' results with lithium are evidently not in order. Extrapolation leads to a boiling point of 2400° K., whereas the directly determined value is 1500° K. The reason may lie in errors in titration; on the other hand, the oven aperture may have been too large for use with a metal so readily contaminated as lithium. It is unfortunate that Bogros omits to give details of the oven dimensions; without them it is not possible to reach a definite verdict.

of no dissociation; and if P_D is the total pressure of the dissociated gas, as measured by a suitable gauge, then

$$\frac{2a}{1+a} = \frac{p_1}{P_D} = 3\cdot41\left(1 - \frac{p_{calc}}{P_D}\right) \quad \ldots\ldots(7\cdot5),$$

from which a can immediately be determined.

This method has been developed by Weide and Bichowsky,[a] and has been used by Bichowsky and Copeland[b] to determine the degree of dissociation of H_2 into atomic hydrogen in a discharge tube; in conjunction with calorimetric measurements on the dissociated gas, they obtained for the heat of formation of molecular hydrogen $105,000 \pm 3500$ cals./mole. Similar measurements have been made by Wrede,[c] who however prefers a static method invented by him to the effusion method as eliminating many of the possibilities of error inherent in the latter. Recently Copeland[d] has determined the heat of formation of oxygen; he finds $131,000 \pm 6000$ cals./mole.

Where, as for example in the case of metallic vapours, the rate of flow N_D and the total pressure P_D cannot readily be determined, it is still possible in favourable cases to arrive at an estimate of the degree of dissociation from measurements of the mass of vapour effusing alone. For

$$\left.\begin{aligned} p_1 &= \frac{2a}{1+a}\cdot P_D \\ p_2 &= \frac{1-a}{1+a}\cdot P_D \end{aligned}\right\} \quad \ldots\ldots(7\cdot6),$$

and ($7\cdot4\,a$) becomes

$$g_D = \frac{P_D a t}{\sqrt{2\pi R T_D}}\cdot\left(\frac{1-a}{1+a}\cdot\sqrt{M} + \frac{2a}{1+a}\cdot\sqrt{\frac{M}{2}}\right)$$
$$\ldots\ldots(7\cdot7),$$

where M is the *molecular* weight; the suffix D refers as before to the fact of dissociation.

[a] Weide and Bichowsky, *J. Amer. Chem. Soc.* **48**, 2529, 1926.
[b] Bichowsky and Copeland, *J. Amer. Chem. Soc.* **50**, 1315, 1928.
[c] Wrede, *Z. Physik*, **54**, 53, 1929.
[d] Copeland, *Phys. Rev.* **36**, 1221, 1930.

Dividing (7·7) by (7·2), which refers to the *undissociated* vapour, we have

$$\frac{g_D}{g} \cdot \sqrt{\frac{T_D}{T}} \cdot \frac{p}{P_D} = \frac{1 + 0\cdot414\alpha}{1 + \alpha} \quad \ldots\ldots(7\cdot8),$$

in which all the quantities on the left-hand side except the ratio p/P_D are directly measurable.

In the favourable case that the temperature range can be extended from negligible to complete dissociation, p/P_D can be obtained by interpolation between the values

$$\frac{p}{p_0} = \frac{g}{g_0} \cdot \sqrt{\frac{T}{T_0}} \quad \text{(dissociation negligible)}$$

and

$$\frac{p}{p_0} = \frac{\sqrt{2}g}{g_0} \cdot \sqrt{\frac{T}{T_0}} \quad \text{(dissociation complete).}$$

The method has been used by Rodebush and de Vries[a] to determine the degree of dissociation of I_2 and (less successfully) of Br_2.

MEASUREMENT OF THE DEGREE OF DISSOCIATION BY BEAM METHODS

The effusion methods just described utilise the properties of molecular *streams* rather than molecular rays proper. There are however ways of studying chemical equilibria which fall strictly under the head of molecular ray methods.

Thus suppose that it is possible to separate out the atoms and molecules in a mixed beam, emerging from a source in which a diatomic vapour is partially dissociated, and to measure the intensities of each separately. Then if α is the degree of dissociation at the temperature of the source

$$\frac{I_\text{atoms}}{I_\text{molecules}} = \frac{2\alpha}{1 - \alpha} \cdot \sqrt{2} \quad \ldots\ldots(7\cdot9).$$

Now it may happen in practice that the detector does not count the numbers of two unlike species on the same basis.

[a] Rodebush and de Vries, *J. Amer. Chem. Soc.* **49**, 656, 1927.

Thus for example the condensation target counts one diatomic molecule as two atoms; so does the surface ionisation detector when the temperature of the tungsten wire (1200° K. to 1500° K.) is high enough to dissociate the molecule into its constituent atoms on the surface.

Let us denote by I'_{X_2} the *measured* intensity of the molecules in a mixed beam of atoms X and molecules X_2. Then for the particular case that $I'_{X_2} = 2I_{X_2}$, we have

$$\frac{I_X}{I'_{X_2}} = \frac{2\alpha}{1 - \alpha} \cdot \tfrac{1}{2} \cdot \sqrt{2} = \frac{\sqrt{2}.\alpha}{1 - \alpha} \qquad \text{......(7·10)}.$$

The *measured* intensities of atoms and molecules referred to the *measured* total intensity I_0' (of the mixed beam) are then respectively

$$\left. \begin{aligned} I_X &= \frac{\sqrt{2}.\alpha}{1 + 0\cdot414\alpha} \cdot I_0' = x I_0' \\[2mm] I'_{X_2} &= \frac{1 - \alpha}{1 + 0\cdot414\alpha} \cdot I_0' = (1 - x)\, I_0' \end{aligned} \right\} \quad \text{...(7·11)}.$$

At least two molecular ray methods are available for the study of chemical equilibria: (1) Velocity Filter Methods, (2) the Magnetic Deflection Method. These we proceed to discuss in detail.

Velocity Filter Methods. The slotted disc velocity selectors described in Chapter 2 may be used to study the degree of dissociation of a diatomic vapour; for if there are two molecular types in temperature equilibrium in the source, the velocities of the molecules of each type will be grouped about a characteristic most probable value α_v; and a plot of the intensity of the beam passing the slotted discs against the speed of rotation would, if the resolution of the apparatus were sufficiently high, reveal two maxima, one for each molecular type.

Actually, the conditions are rather more complicated. Fig. 70 shows for example the intensity distribution to be expected with the Lammert device for the case of a diatomic

vapour dissociated to the extent that the number of atoms
is twice that of the molecules in the beam; a type of detector,
such as for example the condensation target, which counts
one molecule as two atoms is assumed. The full curve I is
drawn, neglecting the width of the parent beam, from the
equation for the *measured* intensity I', namely

$$I'dy = xI_0'.2e^{-y^2}.y^3.dy + (1-x)\,I_0'\,8e^{-2y^2}.y^3.dy$$
$$\dots\dots(7\cdot12),$$

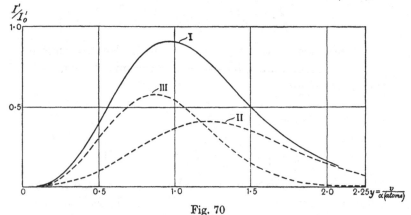

Fig. 70

in which $y = v/\alpha_v$ (atoms), $x = \dfrac{\sqrt{2}\alpha}{1+0\cdot414\alpha}$, for the particular
case that $x = 0\cdot50$. (Compare equation (2·4).) The broken
curves II and III represent the contributions of atoms and
molecules respectively to the summation curve I. It will be
seen that curve I has but a single maximum; x must therefore be evaluated indirectly from the characteristics of the
summation curve, by means of (7·12).

Fig. 71 shows the intensity distribution to be expected
under the same conditions with the Eldridge device. The
summation curve I is here drawn from the equation

$$I'ds = I_0'\left[x.2e^{-(s_a/s)^2}.\frac{s_a^4}{s^5} + (1-x).8e^{-2(s_a/s)^2}.\frac{s_a^4}{s^5}\right]ds$$
$$\dots\dots(7\cdot13),$$

where s_a refers to the atoms, for the case that $x = 0.50$. (Compare equation (2·6).) Again the summation curve has but a single maximum.

The conditions under which it is possible to have two maxima are readily stated. The most general possible type of intensity distribution in the summation curve for a z-atomic

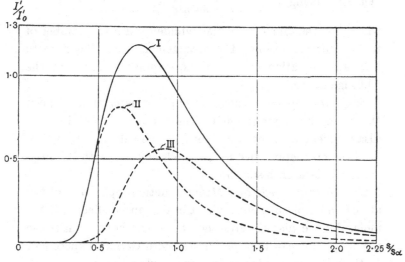

Fig. 71

molecule dissociating directly into its constituent atoms is of the form

$$y^n . e^{-y^2} (1 + A e^{-(z-1) y^2}).$$

Two maxima are first possible when $n > \dfrac{8z}{(z-1)^2}$; for a diatomic molecule therefore when $n > 16$. For the Lammert and Eldridge arrangements n is, as we have seen, 3 and 5 respectively; it is unlikely that an arrangement for which $n = 16$ is a practical possibility.

Zartmann[a] has studied the dissociation of bismuth by means of a velocity filter. A hollow metal cylinder rotating about a horizontal axis was furnished with a single slit in an axial

[a] Zartmann, *Phys. Rev.* **35**, 134, 1930 (Abs.).

direction, which once every revolution passed over a bismuth beam directed vertically upwards. The bismuth atoms or molecules were received on a glass plate, previously sensitised with bismuth, carried on the inner surface of the cylinder opposite the slit. A velocity spectrum was in this way received on the plate, the displacement s for a molecule or atom with velocity v being

$$s = \pi d^2 n/v,$$

where d is the diameter of the cylinder and n the number of revolutions per second. The arrangement leads therefore to the same equation (7·13) for the summation curve as the Eldridge device.

Zartmann reports as the result of photometering the deposit that bismuth vapour at 800° C. is monatomic.[a] This is in sharp disagreement with the results of Leu (see below, p. 187), who used the magnetic deflection method, and also with the spectroscopic data on bismuth.[b]

The Magnetic Deflection Method. Actual separation of the atoms and molecules in a mixed beam can be effected if the molecules are diamagnetic and the atoms have a magnetic moment. One can then perform a Stern-Gerlach experiment on the mixed molecular-atomic beam, when the atoms are deflected right and left of the molecules, which are unaffected by the field.

The first observations of this kind were made on bismuth by Leu and the author, working in Stern's laboratory.[c] It was found that unless the bismuth vapour was sufficiently superheated on its way to the oven slit, in addition to the expected deflected traces a central undeflected trace was obtained, which was ascribed to undissociated diatomic bismuth molecules. Plate VII shows the traces obtained at slit temperatures of 1183° K. and 1428° K. In Fig. 72 a the central

[a] Note added in proof: This conclusion has since been modified by Zartmann himself (*Phys. Rev.* **37**, 382, 1931).

[b] See Grotrian, *Z. Physik*, **18**, 169, 1923; Kopfermann, *ibid.* **21**, 316, 1924.

[c] *Z. Physik*, **49**, 498, 1928 (U. z. M. 8).

undeflected trace is clearly visible; in Fig. 72 *b*, at the higher temperature, the central trace has entirely disappeared.

It was clear that there lay here ready to hand a means of determining the degree of dissociation of a vapour as a function of the temperature; for a comparison of the measured intensities of the central undeflected beam and the total mixed beam would yield at once the quantity

$$(1 - x) = \frac{1 - \alpha}{1 + 0{\cdot}414\alpha}.$$

Preliminary experiments were carried out by Leu,[a] using the oven illustrated in Fig. 73. The oven slit (4) is carried on the extension (14), and is heated by electron bombardment from the glowing tungsten spiral (1), between which and the earthed oven is laid 2000 volts. By this means, the electron energy is concentrated on the slit carrier (27) and there is a consider-

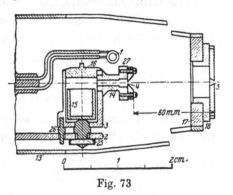

Fig. 73

able temperature drop between the slit and the body of the oven (3), the bottom of which is attached to the water-cooled steel strip (2). It is thus possible to have the substance evaporate from the chamber (15) at a pressure which conserves molecular effusion at the slit, and at the same time to superheat the vapour on its way to the slit through the canal in the extension (14). The temperature difference between slit and oven can be varied by altering the length and diameter of the extension (14). In this way, a range of from about half to almost complete dissociation is covered.

In Leu's experiments the beam was condensed at a nickel target, and a direct quantitative comparison of intensities

[a] Leu, *Z. Physik*, **49**, 504, 1928 (U. z. M. 8).

was therefore not possible. The method used to obtain an estimate of the degree of dissociation was to observe, first the time of appearance t_1 of the trace due to the mixed beam without field; second, the time of appearance t_2 of the central undeflected trace, due to the molecules alone, obtained with field. Then, since the time of appearance may be taken to be inversely proportional to the measured intensity,

$$t_1/t_2 = (1 - x),$$

where x is the fraction of substance present as atoms in the beam.

In evaluating his results, Leu made the error of supposing that x was to be identified with the degree of dissociation α of the molecules in the source; whereas, as we have seen in (7·11), $x = \sqrt{2}\alpha/(1 + 0·414\alpha)$. Thus

$$\frac{t_1}{t_2} = \frac{1 - \alpha}{1 + 0·414\alpha} \qquad \ldots\ldots(7·14).$$

Leu made a further error in setting the pressure p', determined by the effusion method from the equation

$$g = \frac{5·83 \times 10^{-2}}{\sqrt{T}} . \sqrt{209}.p'a \qquad \ldots\ldots(7·14')$$

(at. wt. of bismuth = 209), equal to the total pressure $P = (p_1 + p_2)$; whereas, from (7·4 b),

$$p' = p_1 + \sqrt{2}p_2 \qquad \ldots\ldots(7·15).$$

Leu's value for the heat of dissociation Q

$$(60,000 \pm 15,000 \text{ cals./mole})$$

calculated from the equations

$$\left.\begin{array}{c} \dfrac{\partial \log K_p}{\partial T} = \dfrac{Q}{RT^2} \\[2mm] K_p = \dfrac{4\alpha^2}{1 - \alpha^2} . P \end{array}\right\} \qquad \ldots\ldots(7·16)$$

is therefore incorrect on account of the errors made in evaluating the equilibrium constant K_p.

K_p is correctly evaluated as follows: (7·14) may be written in the form

$$\frac{t_1}{t_2} = \frac{\dfrac{1-\alpha}{1+\alpha}}{\dfrac{1+\alpha(\sqrt{2}-1)}{1+\alpha}} = \frac{\sqrt{2}p_2}{p_1 + \sqrt{2}p_2} \quad \ldots\ldots(7\cdot17),$$

by (7·6). Whence, using (7·15),

$$K_p = \frac{p_1^2}{p_2} = \frac{(t_2 - t_1)^2}{t_1 t_2} \cdot \sqrt{2}p' \quad \ldots\ldots(7\cdot18).$$

Table XII

Temperature T° K.	Time of appearance t_1 of trace without field	Time of appearance t_2 of central trace with field	p' (mm.) from the equation (7·14')	log K_p from (7·18)
	(min.)	(min.)		
1093	13	25	0·57	− 0·453
1093	11	18	0·57	− 0·702
1098	15	30	0·55	− 0·416
1098	11	18	0·55	− 0·717
1148	4	12	0·96	+ 0·263
1153	11	45	0·53	0·237
1173	9	60	0·50	0·526
1263	6	110	0·68	1·194
1263	7	120	0·59	1·100

The values of log K_p calculated from (7·18), using the data of Table XII, are plotted against $1/T$ in Fig. 74. As will be seen the curve is not straight; whether this arises from the possible presence of Bi_n molecules ($n > 2$) in the beam, or from errors in the determination of the times of appearance, some of which are very long, is not certain. The average slope of the curve gives as the corrected value for the heat of dissociation of Bi_2, 26,500 ± 11,600 cals./mole.

The experiments of Leu must of course be regarded as the first tentative steps in a new technique. The method has recently received a notable development at the hands of

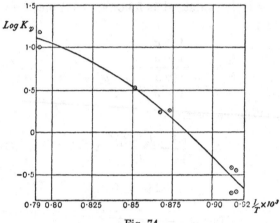

Fig. 74

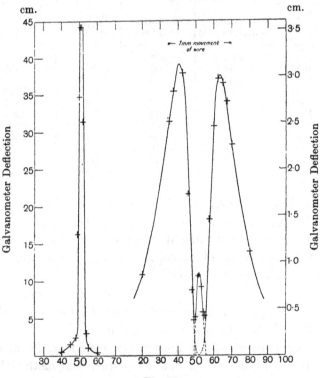

Fig. 75

Lewis,[a] working in Stern's laboratory. He has examined the dissociation of Li_2, Na_2, and K_2 with the surface ionisation detector, with which intensities can be measured to one part in 10,000 of the parent mixed beam. The apparatus was very carefully designed to yield a good vacuum ($\sim 10^{-6}$ mm.), and in particular as constant an oven temperature as possible (temperature variations $< 5°$).

Mr Lewis has very kindly furnished me with the complete data for sodium, from which the power of the method may be judged. Fig. 75 exemplifies the results obtained with sodium at 655° K. On the left is seen the intensity plot of the mixed beam without field. On the right is seen an intensity plot with field. It shows most beautifully how the atoms are magnetically deflected right and left of the non-magnetic molecule beam; the asymmetry in the deflected traces is, of course, that normally present with the Hamburg set-up (see p. 131).

Table XIII

1	2	3	4	5	6	7
Temperature T	$(1 - x)$	ϵ	P	$\log K_p$	$1/T$	Estimated precision of ϵ
(° K.)						(%)
630	0·0165	0·0119	0·104	0·936	1·587	5
655	0·0180	0·0130	0·220	1·223	1·527	2
672	0·0214	0·0155	0·355	1·356	1·488	1
684·5	0·0234	0·0170	0·495	1·459	1·461	1
719	0·0319	0·0233	1·138	1·708	1·391	3–4

Table XIII shows a complete series of measurements for sodium, obtained in a single run; Fig. 75 corresponds to the second entry in the Table. In column 2, $(1 - x)$ is determined directly as the ratio of the measured intensities $\frac{I'_{x_2}}{I_0}$ at the peaks of the corresponding curves. In column 3 is entered

$$\epsilon = \frac{I'_{x_2}}{\sqrt{2}.I_X} = \frac{1 - \alpha}{2\alpha} \text{ by } (7·11) = p_2/p_1 \text{ by } (7·6).$$

[a] Lester C. Lewis, Zeitschrift für Physik, 1931.

The values of the total pressure P in column 4 are those adopted by Ladenburg and Thiele.[a] Now

$$K_p = \frac{p_1^{\,2}}{p_2} = \frac{p_1}{\epsilon} = \frac{1}{\epsilon\,(1 + \epsilon)} \cdot P \qquad \ldots\ldots(7\cdot19),$$

which expresses K_p in terms of ϵ and P. The values of $\log K_p$

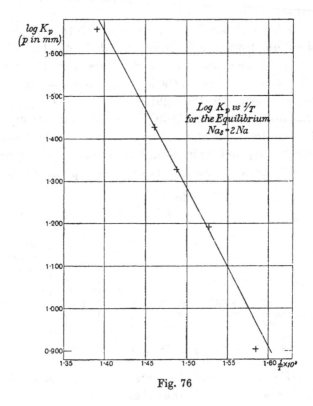

Fig. 76

calculated from (7·19) are entered in column 5, and are shown plotted against $1/T$ (column 6) in Fig. 76. The heat of dissociation as calculated from the slope of the line is 16,300 cals./mole.

    [a] Ladenburg and Thiele, *Z. physikal. Chem.* B, **7**, 161, 1930.

CHEMICAL REACTIONS

Chemical reactions in gases as ordinarily studied are the outcome of one or more of three possible processes: (1) absorption or emission of radiation, (2) intermolecular collisions, (3) reaction at the walls; and it is extremely difficult, and often impossible, to isolate the contributions of the several processes to the total result. Further, it is possible only in isolated favourable cases to gain even an approximate picture of the intermediate steps by which the final products of reaction are attained.

It is in principle possible to use the method of molecular rays both to achieve isolation of the several factors effecting a reaction, and to study the sequence of processes involved with time; the former possibility, thanks to the isolation of the beam in space and to its collision-free character; the latter, to its unidirectional properties.

Three attempts have been made to test the radiation hypothesis, by passing a beam through a furnace; since the beam is practically collision free, any reaction occurring could have been induced by radiation alone. All gave negative results, but all are open to criticism on one score or another. Kröger[a] studied the effect of passing a beam of molecular iodine through a platinum cylinder at 1300° K. Assuming that the iodine molecule breaks up into its constituent atoms under the action of radiation, these atoms are lost to the beam, and the time required to deposit a film of a given thickness on a target placed to receive the beam is increased. Kröger used the colours of thin films as a measure of the deposit thickness. He found the radiation effect to be nil, but undoubtedly underestimated the experimental error in setting it at 1 per cent. The chief weakness of the experiment is however that the dissociation of iodine has never been definitely established as monomolecular.

a Kröger, Z. physikal. Chem. 117, 387, 1925.

Mayer[a] studied the racemisation of pinene in a similar way; a pinene beam was passed through a quartz furnace at round 1000° K., and the pinene deposited from the beam was subjected to polarimetric analysis. The results were entirely negative. The racemisation of pinene is very closely monomolecular, and since no break-up of the molecule is presumably involved, no molecules are lost to the beam. It is however tacitly assumed that the time lag between a possible absorption of radiation and the intramolecular change is small compared with the time of flight of the molecules from furnace to receiver.

Rice, Urey and Washburne[b] studied nitrogen pentoxide, and found the percentage of nitrite in the deposit to be identical with that in the parent beam. This is precisely what would be found in any event whether the radiation were effective in inducing reaction or not; for the products of reaction would be thrown out of the beam and would escape detection.

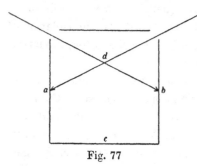

Fig. 77

Kröger attempted also to investigate bimolecular reactions by causing two beams (Cd and I_2, Cd and S_8) to impinge. His results, after disinterment from an almost incredibly verbose account, can be summarised briefly somewhat as follows. The two beams enter a vessel (Fig. 77), crossing at d, and depositing at a and b; products of possible reaction at d are looked for at c. The source pressures used were far too high to retain good beam definition, and molecules scattered from the beam into the region c completely masked any reaction products that might have been present. Next an attempt was made to use the heat of

[a] Mayer, *J. Amer. Chem. Soc.* **49**, 3033, 1927.
[b] F. O. Rice, Urey and Washburn, *ibid.* **50**, 2402, 1928.

formation of (e.g.) CdI_2, which is large compared with the heat of condensation of the reactants, to detect a possible combination of Cd and I_2. Thus the rate of evaporation of liquid oxygen contained in a Dewar vessel surrounding the chamber *acb* was taken to be a measure of the reaction. An increased rate was detected when the two beams entered the chamber simultaneously, but again the results are vitiated by turbulence in the beams, consequent scattering, and hence uncertainty that wall reactions were after all eliminated.

In spite of the negative results so far recorded, the molecular ray method of studying chemical reactions may yet prove to be a valuable one. The initial stages of its development require however a critical approach, careful choice of subjects for investigation, and the patient elaboration of a special technique.

IONISATION POTENTIALS

The standard methods of measuring ionisation potentials V_i give no direct information about the *products* of ionisation. The mass spectrograph evidently furnishes a means of analysis; but there are two essential conditions which must be fulfilled if its statements are to be trustworthy: (1) the mean free path of the ions must be comparable with the distance between the place of their formation and the mass spectrograph; and at the same time (2) the concentration of atoms at the place of formation must be great enough to yield a detectable number of ions. The two conditions are conveniently fulfilled by making the source of the ions a molecular beam.

The method was first used by Smyth,[a] who found that the formation, from a mercury beam, of Hg^{++} ions at 19 ± 2 volts was readily detectable. Ditchburn and Arnot[b] analysed in a similar way the products of the ionisation of potassium by

---

[a] H. D. Smyth, *Proc. Roy. Soc.* **102**, 283, 1922.
[b] Ditchburn and Arnot, *Proc. Roy. Soc.* **123**, 516, 1929.

electron collision (electron energy, 35–100 volts). The ions K_2^+, K^+, K_3^+ and possibly K_3^{++} were detected.

As regards the *ionisation function*, that is the probability N of ionisation of an atom by the impact of a V-volt electron, two independent investigations by a molecular ray method, the first by von Hippel[a] on mercury, the second by Funk[b] on sodium and potassium, agree in assigning a maximum to the NV curve at $V = 2V_i$. There is here a general disagreement with the results of other methods, for which there seems to be no obvious constructional cause. Unfortunately, there is no adequate theory which might serve as arbitrator.

PHOTOIONISATION

The difficulty of avoiding surface effects when using the standard procedure of observing photoionisation in gases has led to the trial of other methods, of which the molecular ray method is one. It is very difficult, however, to assess the value of the method in this field from the results won with it, because these refer to date exclusively to potassium, which substance shows the same anomalous behaviour when examined by any of the current methods. The anomaly is namely this, that in every case a peak in the ionisation curve is obtained for wavelengths round $\lambda 2600$ to $\lambda 2700$, that is, beyond the series limit of $\lambda 2800$. Lawrence,[c] Williamson[d] (the originator of the method), and Ditchburn[e] have all suggested that the $\lambda 2600$–$\lambda 2700$ peak should be referred to the K_2 molecule. The concentration of K_2 in the beam is however undoubtedly very small, and even adopting the interpretation given to his results by Ditchburn—namely, that the probability of

    [a] von Hippel, *Ann. Physik*, **87**, 1035, 1928.

    [b] Funk, *Ann. Physik*, v, **4**, 149, 1930.

    [c] Lawrence, *Phil. Mag.* **50**, 345, 1925. See, however, Lawrence and Edlefsen, *Phys. Rev.* **34**, 1056, 1929.

    [d] Williamson, *Proc. Nat. Acad. Sci.* **14**, 793, 1928.

    [e] Ditchburn, *Proc. Roy. Soc.* **117**, 486, 1928; Ditchburn and Arnot, *ibid.* **123**, 516, 1929.

molecular absorption is some thousand times the atomic—seems insufficient to be the origin of the anomalous peak.

Ditchburn and Arnot (*loc. cit.*) analysed the products of the photoionisation of a potassium beam with a mass spectrograph, and obtained only the K^+ ion; this does not however exclude the possibility that the primary product is K_2^+, since the K_2^+ ion is demonstrably unstable, and can split up according to the scheme

$$K_2^+ \rightarrow K^+ + K.$$

The technique of the production of the beam in all the experiments cited was comparatively crude, and could in any case yield only comparative values of the absorption coefficients. However, the beam method does not appear to be inherently merely a comparative one;[a] where a quantitative measurement of the beam intensity is possible, the absolute determination of atomic absorption coefficients is not excluded.

HYPERFINE STRUCTURE OF SPECTRAL LINES

The unidirectional, collision-free properties of a molecular beam have been utilised in the study of the hyperfine structure of spectral lines.

The natural width of spectral lines is increased by a number of causes: by the magnetic or electric influence of neighbouring molecules, by collisions, and by the Doppler effect arising from the temperature motion of the radiating molecules towards or away from the observer. In a molecular beam, broadening of the lines from the first two causes is negligible, and if the beam is observed at right angles to its direction, the component of the velocity of the molecule in the direction of observation is so small that the Doppler broadening is reduced to a value corresponding to a temperature of only a few degrees absolute. Thus spectral lines emitted from a molecular

[a] See for example Mohler, *Reviews of Modern Physics*, 1, 219, 1929.

beam as source are extremely narrow, and therefore very suitable for the study of minute structural details.

An atomic beam can be excited to emit the resonance lines, as was first shown by Dunoyer[a] for sodium. Dobrezov and Terenin[b] have used a sodium beam excited to emit the D-lines to study their hyperfine structure, and have been able to observe each D-line as a very close pair. Their results are in excellent quantitative agreement with those of Schüler,[c] obtained with the hollow cathode method.

Bogros[d] has found that a lithium beam can be excited to resonance by $\lambda 6708$. Recently[e] he has analysed this line, which has of course long been known to be a very narrow doublet, in absorption at a lithium beam. He finds a doublet separated by $0\cdot15$ Å., and a weak satellite, which had previously been observed but not measured by Schüler and Wurm,[f] $0\cdot156$ Å. towards the red. The satellite is probably the long wavelength component of the Li (6) doublet, the short wavelength component of which falls on the long wavelength component of the stronger Li (7) doublet.

Dr Minkowski informs me that he has made preliminary experiments on the excitation of coarse molecular beams by electron collision. He has obtained in this way sufficient light intensity to encourage the hope that the "molecular ray lamp" may be applied to the study of the hyperfine structure, not only of resonance lines, but of spectral lines in general.

[a] Dunoyer, *Compt. rend.* **157**, 1068, 1913; *Le Radium*, **10**, 400, 1913.
[b] Dobrezov and Terenin, *Naturwiss.* **33**, 656, 1928.
[c] Schüler, *Naturwiss.* **16**, 512, 1928.
[d] Bogros, *Compt. rend.* **183**, 124, 1926.
[e] *Ibid.* **190**, 1185, 1930.
[f] Schüler and Wurm, *Naturwiss.* **15**, 971, 1927.

APPENDIX

The velocities of the molecules of a gas are distributed about a most probable value α according to the relation

$$dn_v = \frac{4n}{\alpha^3 \sqrt{\pi}} \cdot e^{-v^2/\alpha^2} \cdot v^2 \cdot dv \qquad \ldots\ldots(1'),$$

where dn is the fraction of the number n of molecules in unit volume which have velocities between v and $v + dv$.

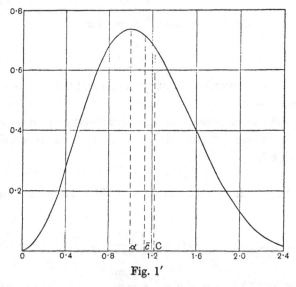

Fig. 1'

Equation (1') is exhibited graphically in Fig. 1'. It is at once clear that the curve is not symmetrical about the α-ordinate, but that the area under the curve is greater on the high velocity side. The *average* velocity $\bar{c}$ will then be greater than the most probable velocity α. In fact

$$\bar{c} = \frac{\displaystyle\int_0^\infty v \cdot dn_v}{\displaystyle\int_0^\infty dn_v} = \frac{2}{\sqrt{\pi}} \cdot \alpha \qquad \ldots\ldots(2').$$

The kinetic energy of a given molecule is $\frac{1}{2}mv^2$; the mean kinetic energy of a number n of molecules is therefore $1/2 . nm . \overline{v^2} = 1/2 . nm . C^2$. If the Maxwell distribution of velocities (1′) be assumed, then

$$C^2 = \frac{\displaystyle\int_0^\infty v^2 . dn_v}{\displaystyle\int_0^\infty dn_v} = \tfrac{3}{2} . a^2,$$

and the *root mean square* velocity

$$C = \sqrt{\tfrac{3}{2}} . a \qquad\qquad \ldots\ldots(3').$$

From (2′) and (3′) we have the relation

$$\bar{c} = \sqrt{\frac{8}{3\pi}} . C \qquad\qquad \ldots\ldots(4').$$

It is often convenient to express a, $\bar{c}$ and C in terms of the temperature.

Thus the mean kinetic energy of translation of a molecule is, since to each degree of freedom must be assigned the energy $\frac{1}{2}kT$,

$$\tfrac{1}{2}mC^2 = \tfrac{3}{2}kT,$$

where k is Boltzmann's constant and T the absolute temperature.

Therefore $\qquad\qquad C = \sqrt{\dfrac{3kT}{m}} \qquad\qquad \ldots\ldots(5'),$

whence by (3′) $\qquad\quad a = \sqrt{\dfrac{2kT}{m}} \qquad\qquad \ldots\ldots(6'),$

and by (4′) $\qquad\qquad \bar{c} = \sqrt{\dfrac{8kT}{\pi m}} \qquad\qquad \ldots\ldots(7').$

AUTHOR INDEX

SUBJECT INDEX

Printed in the United States
By Bookmasters